SCIENCE AND TECHNOLOGY
FOR DEVELOPMENT

Science and Technology for Development

Corporate and Government Policies and Practices

Jack N. Behrman
University of North Carolina

William A. Fischer
University of North Carolina

A project sponsored
and administered by the

FMME
**Fund for Multinational
Management Education**

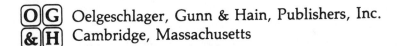 Oelgeschlager, Gunn & Hain, Publishers, Inc.
Cambridge, Massachusetts

This book was prepared with the support of NSF Grant PRA77-21851. However, any opinions, findings, conclusions and/or recommendations herein are those of the authors and do not necessarily reflect the views of NSF.

International Standard Book Number: 0-89946-023-2

Library of Congress Catalog Card Number: 79-27213

Printed in the United States of America

Library of Congress Cataloging in Publication Data

Behrman, Jack N
  Science and technology for development.

  Includes index.
  1. Research. 2. Underdeveloped areas—Research. 3. Underdeveloped areas—Science and state. 4. Science and state—United States. 5. International business enterprises—Research.
  I. Fischer, William A., joint author.
  II. Title.
  Q180.A1B394   338.9   79-27213
  ISBN 0-89946-023-2

Contents

Preface

Both of us have for some years been interested in the role of technology in industrial development, one of us considering the international aspects, the other domestic implications for productivity and innovation. As a result of both several forays into the problems of technology transfer abroad and repeated conversations with government officials, it became clear to us in 1976 that the next major area of business-government and government-government dialogues would be on the location of research and development (R&D) activity.

One of us made some effort to determine the status of international R&D activities, only to find that the literature provided little information. Consequently, it seemed necessary to study R&D activities abroad. However, information about R&D activities overseas was too sparse even for designing a good study. Therefore, we decided to undertake some case studies of companies to elicit the necessary information to design a more extensive investigation. Fortunately, Unilever's Director of Public Affairs (Doreen Wedderburn) quickly saw the desirability of greater public knowledge of the R&D basis for international activities and obtained corporate approval for an intensive study of R&D behind some of Unilever's worldwide product lines. We decided to compare several companies' activities, rather than try to detail all of one company's R&D around the world. We considered the soap and detergent line of Unilever comprehensible to a nonscientist—at least sufficiently for the purposes at hand. These purposes were to determine why and how R&D activities were set up abroad and how they were managed and coordinated—and what infrastructure was needed—so as to learn the lessons for appropriate policies toward R&D in developing countries.

Once Unilever's study was underway, we obtained agreement from four other companies: Pfizer, Johnson & Johnson, Du Pont, and General Motors. We are very grateful to the individuals in each company who obtained permission for the case studies on an attribution basis (which is not always easy for companies to agree to). Equally, we are greatly indebted to the many company officials who gave unstintingly of their time in interviews and in gathering and forwarding information from files.

These case studies were conducted during the years 1977–1978, frequently with one of the R&D officials of the company going along on trips abroad so as to assure full cooperation and to help "translate" company and scientific communication. Without this assistance, some of the studies would not have been so readily passed through the review process, which was required to remove any confidential information that might have crept in and to correct misinterpretations. (In fact, only a few sentences out of several hundred pages were censored for confidentiality; in some cases, material that might have been considered sensitive was left in to make the story more complete.)

Much of the data in the studies became out of date as we moved toward publication. The companies we were working with appropriately questioned whether they should update their materials. However, to do so would have been impossible as the target moved too rapidly. It was therefore decided to present the cases as snapshots of the R&D activities being conducted during one period of time, noting that none of the companies still look as described here. The reader should bear in mind that our objective was not to paint pictures of particular companies; rather, it was to find out from them what were the processes, procedures, problems, and patterns of R&D activities overseas so that we could apply these lessons at least tentatively to policy issues. Because of this limited purpose and the volume of material drawn from these five studies (running to over six hundred typed pages), we decided to publish only three, as summarized in Chapter 1, using the others to buttress the expanded survey.

The results of the five studies were sufficiently interesting and useful that officials of the National Science Foundation (NSF) decided to test them in a larger sample. During the summer of 1978 Professor Fischer was brought into the project to help conduct these interviews. He focused on the U.S. companies while Professor Behrman interviewed a number of European and Japanese companies.

Through former contacts with companies in the United States and abroad, as a result of research on technology transfers and foreign investment, and with the irreplaceable assistance of the Industrial Research Institute (IRI) in the United States, we obtained the cooperation of 34 companies in the U.S., 16 in Europe, and 6 in Japan. Each agreed to day-long interviews at headquarters, replying to questions springing from the case studies. The IRI helped us in determining which companies were likely to have experience overseas in R&D activities and in selecting a fairly representative industry sample. Again we are indebted to numerous individuals (sometimes more than one per company) who were interviewed without attribution as to company or official and who gave us the information summarized in Chapter 1.

While the case studies were under review, Professor Behrman was offered an opportunity to study the extension of pharmaceutical R&D into developing countries in the area of tropical diseases. This study added new dimensions and carried the investigation into several developing countries, with the assistance of officials from a dozen international pharmaceutical companies and officials of the International Federation of Pharmaceutical Manufacturers Associations (IFPMA). Some of the materials gathered in this study are attributed to the companies involved; others that came from the personal experiences of officials were left unattributed. The results, summarized in Chapter 2, dovetail with those from the other two stages in expanding the view of the route to innovation and commercialization, though, obviously in a unique field.

To complete the picture of R&D problems in developing countries (LDCs), the NSF agreed to a further study of the science and technology policies in LDCs as they related to industrial R&D, and the role that was being and could be played by international companies. This study was conducted during the summer of 1978 and the winter of 1978/79. It covered 9 LDCs and encompassed interviews with 75 science and technology (S&T) officials. The results are summarized in Chapter 3.

These studies permitted some conclusions to be drawn for recommendations to the U.S. government in its preparation for the U.N.

Conference on Science and Technology in Vienna (August 1979). It was for this purpose that the final chapter in this monograph was written. It is couched in terms of potential follow-up to the Action Program of the conference, which called for increased support of S&T policies and programs in developing countries.

The studies form a package of investigations into R&D activities overseas by international companies, special problems in one area of pharmaceutical R&D, S&T policy issues in LDCs, and alternative bilateral programs for support of S&T activities in LDCs and R&D contributions of the international companies. The study package is drawn together in this monograph, though the larger studies summarized in Chapters 1 through 3 have been published separately.

It is hoped that the many individuals, corporate managers and officials at several levels in foreign governments and in the U.S. government who contributed so much time to these investigations and the members of the Fund for Multinational Management Education who sponsored all the studies (along with the NSF and several contributing companies) will consider the results worthwhile and will be able to use them in the furtherance of their mutual goals to employ science and technology effectively to achieve desired industrialization.

Jack N. Behrman
William A. Fischer
Chapel Hill, N.C.
November 1979

Introduction: Purposes and Methodology of Studies

The developing countries have for a long time been concerned about their technological dependence on the advanced countries. Consequently they have recently begun to examine means of developing their own science and technology (S&T) capabilities. Some of them have already moved fairly far in this direction while others are making first steps. In line with the view that much of the S&T development comes from the large national corporations, their secondary concern has been the role that transnational corporations (TNCs) might play either as potential roadblocks or contributors to the development of indigenous capabilities in LDCs. It was clear some years ago that these concerns would become one of the subjects of the North-South dialogue.

To provide government and company officials with more adequate information on the patterns of R&D activities and the development of S&T policies, our research was designed to elicit data on what was taking place and what the basic impediments might be to the stimulation of S&T for international development.

Since there was very little literature on the policies and practices in overseas R&D activities of TNCs, the first step in the research was to find several companies that would cooperate in the development of fairly extensive cases of their activities. Five companies agreed to cooperate—Unilever, Pfizer, Johnson & Johnson, Du Pont, and General Motors—permitting extensive interviews with their officials both at headquarters and in their affiliate labs. Since it was impossible to describe the entire R&D activities of such companies, specific product areas and geographic locations were selected in consultation with company officials, so as to provide insights, rather than complete descriptions, into company policies, practices, and problems. The criteria were that the product area selected not be so extensive or complex that it could not be adequately described, that it have labs overseas working on related projects, and that its managers were willing to cooperate in the study. Visits were made to R&D centers of each of the parent companies to obtain a description of their facilities and programs and management practices. Overseas labs were visited to determine the reason for the geographic location as well as lab organization, types of facilities developed, staffing patterns, budgets, capabilities, project selection and management, coordination with other projects throughout the company, and application of the results of the various projects. We examined this overseas experience particularly from the standpoint of determining its application to future developments in the LDCs. In addition, discussions elecited the decision criteria for the location of the labs, reliance on the local science community, governmental incentives and restrictions, and so on. We examined any cooperative R&D activities by the overseas labs, and we determined the process of diffusion of science and technology through contacts in the local industrial community and mobility of personnel. Finally, we discussed assistance by the companies to the gestation of an indigenous S&T capability.

We then returned these case studies to the companies for verification of their accuracy and suitability for publication. Despite the fact that the respondents were exceedingly candid, only a small portion of the material was deleted to protect confidentiality. Company concern over inaccuracies stemmed largely from the fact that the situation in each case changed as the study progressed. Agreement was reached, however, to record the activities as they were when studied as it would have been impossible to keep updating the materials. Therefore these studies describe *past* practices though many of the same policy issues remain.

To elicit full responses from officials, several companies sent top-level R&D officials of the company with the interviewer; he could therefore assure the respondent that he should answer fully and candidly, and he was also able, as the interviews proceeded, to clarify both questions and answers and later on assure those reading the drafts that the materials included were representative of the conversations. Even so, it must be emphasized that the materials included throughout all parts of this research are perceptions of respondents and therefore constitute anecdotal evidence. Given the small number of companies visited and officials interviewed, no applicable conclusion could be drawn from the cases. Nor was this the purpose of the research. Rather, it was to first develop information that would be useful in designing a more adequate research project and then to carry that second stage forward with a larger number of companies, subsequently doing some in-depth studies of very specific activities—in this case, pharmaceutical research in tropical diseases—and finally to apply the lessons learned to future U.S. policies toward S&T in development.

One research objective was to provide inputs into the formation of U.S. policies in preparation for the U.N. Conference on Science and Technology in Development (UNCSTD) in Vienna, August 1979, and for subsequent dialogues with companies and governments after that conference. Therefore the entire research is policy oriented—that is, to the policies of companies, the U.S. government, and developing-country governments. For this purpose, the research objective was to identify the policy issues that would arise out of the concerns of governments, the objectives of companies, and the limitation on each.

Given the results of the case studies, it was possible to design a set of more specific interview questions that focused on the major policies and practices of a larger group of companies. The interviews were conducted with 56 U.S., European, and Japanese international companies. They were made possible through the generous help of not only corporate officials but also officials of the Industrial Research Institute in the United States and its members and affiliates in the United States and Europe. The interviews with company officials were open-ended, but they started from a structured set of questions about overseas R&D activities, locations of the labs and criteria for their location, motivations for their establishment and their growth paths; management styles and methods of coordination; patterns of invention and innovation; collaborative R&D activities; process of diffusion of R&D capabilities from TNCs into host communities; and the relationships between company policies and host-government incentives or inducements.

From these interviews a few generalizations could be drawn, but it became clear that much of the R&D experience abroad was strongly affected by the industry sector and management style of the company. These differences among companies must be taken into account in governmental policy formation if desired results are to be achieved. In addition, it became clear that company decision-making was significantly affected by the environment within which R&D activities would take place. Research and development is merely one piece of a complex continuum of activities, demonstrated particularly well in the case of pharmaceuticals, in which the market is of a state origin.

The complexity of the interlinked aspects of invention, innovation, and commercialization were studied in some detail in the R&D activities of pharmaceutical companies in the area of tropical diseases. Many questions had focused on the pharmaceutical companies on the part of the World Health Organization and members of the U.S. Congress as to whether or not they are doing enough to seek out solutions to critical tropical diseases. To find out what in fact was being done to mitigate any of the diseases, research was designed to examine each of the phases from basic research to commercialization for delivery of the drugs. This phase of the project required interviewing a number of the major pharmaceutical companies in the United States and Europe, particularly their lab directors, and following the application of the results through clinical test stages and field trials and into delivery systems such as through public health institutes. The results of this study—though uniquely applicable to the pharmaceutical industry—point out the close linkages required among corporate, university, and government institutions in order to apply research results. Policies formulated in any one of these sectors require cooperative policy positions on the part of the others.

To complete the assessment of S&T contributions to development, the State Department asked us to examine the government policies of nine developing countries: Brazil, Egypt, India, Indonesia, Iran, South Korea, Malaysia, Mexico, and the Philippines. These were chosen not only for their geographic dispersion but also because of their efforts to strengthen their S&T communities and their significant roles in preparations for the UNCSTD in Vienna. Interviews with appropriate officials were arranged by U.S. Embassy personnel at each post, with an original intention to focus on the relationship of domestic S&T policies to the attraction of R&D activities of TNCs. However, it was soon clear that governmental consideration of the contributions of TNCs was minor and that the study had to focus principally on the development of indigenous capabilities with only peripheral attention given to governmental policies toward the TNCs. However, attention was given to the relationship of domestic

S&T policies toward R&D activities of local companies, noting incentives or inducements, which would also be applicable in the main to the TNC affiliates.

In this part of the investigation, therefore, primary attention was given to the process of gestation of S&T policies and practices within the country, beginning with the formation of economic and social goals that S&T policies were to further. Attention was also given to the organization and structure of decisionmaking within the government, which would buttress S&T policy, the impact of a colonial heritage, and future objectives related to S&T policies. The principal thrust of the examination of LDC policies was to describe their progress toward the formation of the "research triangle" composed of government policies and institutions, university and academic contributions, and industrial R&D development and application. Special attention was given to the formation of government institutions for R&D—whether independent or tied to ministries or state enterprises. Company roles were examined, both local and foreign-owned affiliates. And questions were directed toward the existence of linkages among the government institutes, university laboratories, and public or private enterprises. Finally, inquiries were made as to future needs of each of the host countries in accelerating the development of indigenous S&T capabilities. Since the interviews were open-ended, questions were addressed to those deemed most capable of responding appropriately; it would have wasted valuable time to structure interviews just to gain statistical data.

Again, as with the companies, it was impossible to do a complete survey of the S&T policies and activities of each of the countries. Such an effort would have meant a book for each of the countries themselves. Instead, the objective was to illuminate the process of S&T gestation, the obstacles encountered, and prerequisites, and to determine whether or not there were any key junctures at which advanced country governments could be of assistance or where the TNCs could make singular contributions. The responses showed the nine countries to be in several positions along a continuum from tentative steps toward formation of an S&T community to a carefully structured program including clear policies, effective R&D institutions, and useful linkages with industry.

The results of these studies led to an appreciation of the capabilities and limitations of the U.S. government in its policy responses to the needs for S&T in international development, which led to the conclusion that the primary effort on the part of the United States should be bilateral in nature—and targeted at specific needs in each

LDC where a program is mounted. In addition, an effective U.S. program requires heavy reliance on implementation by the private sector—that is, TNCs and labor unions.

Each chapter in this monograph summarizes the content and conclusions of the major studies just described. They reflect investigations of different aspects of the S&T process as seen from different participants in it. All types of participants have been interviewed, but not all of those engaged in any one activity. Different perceptions would probably be held by other participants in the process. Therefore, this study does not purport to be complete or wholly accurate. Its accuracy is buttressed by the fact that we were able to go back to some of the participants for checking, and in some cases we were able to cross-check from one respondent to another. As much of this verification was done as was feasible within the time limits of the research. The information base should therefore not be considered as definitive but more as suggestive of the policy orientations and problems surfaced during the research.

Even so, except for Chapter 4, "U.S. Responses," written by Professor Behrman, the results of the studies are based on the views of others rather than on those of the authors themselves. It is in the drawing of lessons and conclusions that we have injected our own assessments.

One of the lessons we have drawn from the studies is that much of the information on international R&D activities and S&T policies is inadequate and tends to lead to erroneous policy conclusions simply because company officials, R&D managers, and government officials are making numerous trade-offs that are not explicit in policy formation because of an inadequate understanding of either the R&D process by companies or the S&T process in international development. This is not to say that we have described the processes fully or adequately—only that we have learned enough to know that what was known before was likely to lead to inappropriate policies from the standpoint of the pursuit of the expressed goals and objectives.

Therefore further research is needed, looking to an elaboration of the relationships among the various elements of the process, that would lead to a better understanding of the trade-offs that are made—for example, among ownership, control, technology transfers, and location of R&D activities among affiliates of TNCs. With an understanding of these trade-offs, government policies attempting to provide incentives or negative inducements to location of such activities would be better informed. At present, many government policies are developed as though trade-offs do not occur among different aspects of company policies and practices. Thus governments assume that regulations requiring minority ownership by TNCs will not have an adverse effect on technology flows or the establishment of R&D laboratories; they are likely to in both ways since control over the "heart" of the business is significantly reduced on the part of the technology generator or supplier.

These and other lessons are noted throughout the study, and many of the views will signal new areas of research for others—such as the comment by the Mexican government that state R&D institutions appear to be more readily formed and more useful in areas where the technology moves slowly and is fairly unsophisticated, permitting the institution to make modifications in technology applicable to the host country as compared to more rapidly changing technologies and highly sophisticated, which require dedication of considerably larger capital and manpower resources that are not readily available to developing countries.

Finally, if further research demonstrates that there is a fairly logical and repeated process in the gestation of S&T for development, it will become easier to determine the junctures at which foreign assistance (public or private) could be effective. The present studies at least provide a framework for subsequent studies to elaborate through an examination of the patterns of development within various industry sectors and among different developing countries.

Chapter 1

Overseas R&D of International Companies

In an increasingly technological age, the attainment of economic growth requires the establishment and nourishment of indigenous scientific and technical resources. The transnational corporations (TNCs) represent a particularly important source of such scientific and technical capability. In dialogues preparatory to the U.N. Conference in 1979, considerable attention was directed by government officials to the conditions under which TNCs offered technology and their willingness to promote technology in the developing countries. It was the purpose of this study to develop, through field interviews with R&D executives in American, European, and Japanese transnational corporations, an understanding of the decision processes under which these corporations perform R&D abroad, choose a site for foreign R&D activities, manage their foreign R&D activities, make a contribution to foreign scientific and technical capabilities, and react to foreign government pressures. Such information was sought to assist policymakers in TNCs themselves and the host- and home-country governments to understand and influence the transfer of scientific and technical capabilities.

These interviews began in 1977 with the case studies summarized on pp. 28–56 below and were followed by the wider survey recorded

All the materials summarized here are taken from our study *Overseas R&D Activities of Transnational Corporations* (Cambridge, Mass.: Oelgeschlager, Gunn & Hain, 1980).

on pp. 10–28. Since a major thrust of this study was toward problems in LDCs, the third section reviews in depth one company's assessment of its future role in R&D in developing countries. Other companies may have had similar views—and some did express agreement on some of the points—but Unilever was the only one with which this subject was pursued so intensively. It was questioned closely because its R&D activities were so clearly structured in LDCs and followed a precise and well-communicated policy, starting with embryonic developmental efforts and moving up the scale of R&D complexity. Whereas R&D activities in advanced countries abroad are often set up full-blown, either from acquisition or establishment, they almost never are in an LDC.

SURVEY OF SELECTED COMPANIES

During the spring and summer of 1978, following the five case studies, interviews were conducted with R&D executives in 34 American, 16 European, and 6 Japanese firms. The interviews used a structured format and covered a variety of topics, including: the location decision; the scope of R&D conducted abroad; entry modes and growth paths of foreign R&D activities; the management, funding, and coordination of foreign R&D activities; the diffusion of R&D-related capabilities to host-country scientific and technical communities; and the relationship between the transnational corporations and foreign governments.

For the most part, the firms interviewed were known beforehand to have foreign R&D activities. In addition, they were all members of the Industrial Research Institute in the United States, the European Industrial Research Managers Association, or had prior contacts with the researchers so that their research intensiveness was known beforehand—four companies in the United States had no R&D overseas, but they were included to determine why not since they were research intensive in the United States. Even with this bias toward firms willing to perform R&D abroad, the results of the interviews provide an emphatic rejoinder to the familiar comment that "the transnational corporations aren't doing R&D abroad." In fact, even the more specific observation that "the transnationals aren't doing any R&D in the developing world" was found to be incorrect. However, the purpose of the study was *not* to discover *how many* companies were doing R&D abroad but rather to ascertain in those cases where the firms were pursuing R&D abroad, why they had set them up and how they managed those activities.

Location and Scope of the Foreign R&D Activities of Transnational Corporations

The interviews conducted for this study clearly indicated that there is considerable R&D activity by transnational corporations in locations other than their own parent countries. Of the 34 American transnationals interviewed, 31 reported having some overseas R&D activities. Among these, we were able to identify 106 active foreign R&D groups. Furthermore, the European transnationals appeared even more active in the pursuit of R&D abroad, with the 16 firms interviewed reporting 100 distinct foreign R&D activities. While more of these foreign activities were in fact smaller in size and more restricted in scope than were the R&D activities pursued at home, they were clearly involved with R&D and not simply technical services or quality control. These numbers represent a commitment to R&D abroad by the transnational corporations interviewed that far surpasses expectations derived from a reading of the rather sparse literature concerning the foreign R&D activities of TNCs.

Typically, the foreign locations chosen by the TNCs for their R&D activities were dominated by the advanced industrialized nations. As an example, almost one third of the American firms had R&D activities in the United Kingdom, Australia, Canada, and Japan, while nearly all the European firms had R&D activities in the United States, and more than one half had R&D activities in France. While these findings agree with previous studies, there was more variety in locations than might have been expected. More than 30 foreign countries hosted R&D activities of the firms interviewed, and several TNCs had located R&D activities in advanced developing countries, the most popular of which were Brazil, India, and Mexico (see Table 1.1).

The majority of the R&D being performed abroad by TNCs is primarily addressed to *applied* research and development. However, this varied considerably, depending on market orientation. About a quarter of the foreign R&D activities identified by American firms had missions that included a substantial commitment to new produce research, and at least four of those were expressly concerned with exploratory (basic) research. As might be expected, all the foreign laboratories of American firms that had new product research responsibilities were located in Europe, Japan, or Canada, and in one instance Australia. European firms appear to differ from their American counterparts with regard to the performance of R&D abroad in two important aspects: First, the European firms appear to be considerably more willing to assign new product

Table 1.1   Country of Location of Foreign R&D Activities of 31 American and 17 European Firms

Country	Number of Firms with R&D Activities Located in Each Country[a]	
	American	*European*
United States		14
United Kingdom	11	5
Australia	8	0
Canada	8	1
Japan	8	4
France	7	10
Germany	6	9
Mexico	6	1
Brazil	5	5
Netherlands	4	1
Switzerland	3	1
Hong Kong	2	1
India	2	6
Italy	2	4
Spain	2	4
Argentina	1	2
Austria	1	2
Belgium	1	1
Colombia	1	0
Denmark	1	0
Ecuador	1	0
Egypt	1	0
Greece	0	1
New Zealand	1	0
Norway	1	0
Philippines	1	0
Singapore	0	1
South Africa	1	1
Sweden	1	4
Taiwan	1	0

Source: Interviews.
[a]Multiple labs in a country are counted as one location.

development responsibility to their foreign R&D activities than are American firms; and second, European firms are more willing to pursue new product research in the advanced developing countries. The six Japanese firms interviewed—in chemicals, electronics, heavy machinery, petrochemicals, and construction—had not yet established any foreign R&D activities, and they reported that no others had to their knowledge.

The Relationship Between the Market Orientation of Transnational Corporations and Their R&D Activities Abroad

A central theme running through almost all our interviews on the three continents was that TNCs would probably prefer to do their R&D in one centralized location, if they possibly could. A variety of factors, however, arise that lead them to consider R&D activities at other locations. A fundamental determinant of the relative influence of these factors is the market orientation of the firm. The firms most adamant about centralizing their R&D activities are those primarily interested in serving only their domestic market. To the extent, however, that these firms needed to have operations abroad for extracting raw materials or performing assembly operations, they also came to need technical service functions abroad. Often these technical assistance activities evolved into R&D groups with some process-related responsibilities.

Seven firms in our survey, all American, were categorized as having a "home-market" orientation—that is, seeking investments in the extractive sector to bring materials home. Typically, they had little or no sales in foreign markets. In those few instances where they did sell to foreign customers, they viewed such activities as being direct extensions of their domestic business and not requiring any further R&D beyond that which had been performed for the original, domestic market. Because of their extractive operations these firms tend to have high exposure in the developing world. They have not, however, located much R&D activity in these countries because they typically do not refine raw materials or make finished products in these markets.

Among these seven home-market firms, there were only two foreign R&D activities identified, and both of these originated as technical assistance activities in support of a foreign extractive operation. There was never any intention of these laboratories becoming significant R&D performers in their own right, and neither of them had new product or process responsibility.

Firms involved in the marketing of goods and services designed to satisfy primarily local styles and tastes have a compelling reason to do R&D abroad. These firms need to be as close to their markets as possible and are designated as "host-market" companies. Industries typified by companies with this market orientation span a wide range of technical sophistication from chemicals and pharmaceuticals to foods and textiles. The products of the host-market firms typically exhibit a high degree of standardization within any national market but not necessarily between markets. There were 23 such firms among the American companies in our sample—7 pharmaceutical, 4 chemical, and 4 food and tobacco firms—plus 15 European firms.

Of the foreign R&D activities, 91 percent identified by the American TNCs belonged to host-market firms. In most cases, the decision to go abroad and the location selected reflect important market-related considerations. Illustrative of these considerations is the agricultural chemical firm that needs to test its products in the markets they are intended for and hence has facilities in South America to treat South American pest problems, facilities in the Far East to address problems of tropical climates, and facilities in the Philippines to provide market conditions indicative of Japan. Similarly, all the American pharmaceutical firms interviewed had European formulation laboratories as a result of European regulations and preferences for drug dosages and practices that differ from those in the United States.

Host-market firms also tended to endorse the proposition that their foreign affiliates are often serving distinctive markets and, as such, are autonomous business entities. Accordingly, the primary initiative for the establishment of foreign R&D activities in these firms often arose within the foreign affiliate itself. In such cases, it was not unusual for the affiliate to make the decision to establish an R&D group when it felt that conditions were "right." Such R&D groups typically focused on applied R&D projects.

While all the firms interviewed expressed an interest in enhancing their relationship with host-country governments, the host-market firms, with their distinctly national-market focus, were particularly sensitive to the importance of such relations. In a number of cases, this sensitivity resulted in the establishment of foreign R&D activities, some of which have become particularly productive.

Unlike the firms with home-market orientations, the host-market firms appear to be more willing to assign new product research responsibilities to their foreign R&D activities. Of the foreign R&D laboratories, 26 percent of the host-market firms had such a mission. This undoubtedly reflects the wide diversity in business conditions faced by these firms in their different market. Such diversity was characterized by an executive in a food company when he observed that "in each market we must work with different raw materials to make products satisfying different tastes, while at the same time meeting different government standards."

Five of the firms in our sample were engaged in diverse sectors but produced items that were standardized the world-over and that reflected extremely high scientific and technical achievement.[1] Such firms are characterized as "world-market" oriented. They became involved with R&D abroad in an effort to acquire the best international scientific and technical talent they could get, irrespective of its location. Unlike the firms with either of the other two market orientations, these world-market firms typically established their R&D abroad without regard for the location of their existing international operations. They were much more attracted to the concentration of knowledge and talent than to market size. Once established, the foreign R&D activities of these firms often became world leaders in their particular technical fields. But, given their world-wide, standardized market, there is little *need* to diversify R&D geographically. Therefore these firms had few foreign R&D locations.

Of the seven foreign laboratories identified by these world-market firms, five had new product research missions, and three of these were dedicated to research of an exploratory nature. When locating R&D abroad, these firms are primarily concerned with the local availability of specific types of skills in particular technical areas. As one executive put it, "The key to our foreign laboratory's success is the building up of several good quality technical areas, above threshold size to the point where they can be self-sufficient and yet still engage in cooperative research." The development of such capabilities abroad also facilitates the firm's access to foreign scientific and technical communities. This was noted by an American R&D manager who observed that his firm's Swiss laboratory ". . . is really a European lab, and through it we gain access to all of Europe."

Market Orientation and the Establishment and Growth of R&D Abroad

The foreign R&D activities of TNCs arise in one of three ways:

1. Evolution from foreign manufacturing or market-service operations, either wholly or jointly owned.

2. Direct placement of R&D activities in a foreign location as an extension of the parent company's R&D program or as a cooperative effort between two or more companies.
3. Acquisition of existing foreign R&D activities.

The market orientation of a firm, which strongly influences its decision to perform R&D abroad and its choice of foreign R&D locations, also is an important determinant of the means by which the firm establishes and develops its R&D activities abroad. In general, however, the more lengthy evolutionary process from technical service activities to R&D is likely to be associated with groups pursuing applied R&D, while the more expeditious direct placement is associated with the establishment of foreign R&D groups whose mission includes new product responsibility. Acquisition of existing foreignR&D laboratories accounted for about 25 percent of the R&D activities in either group.

Evolution. In slightly more than one half of foreign R&D activities of the American transnationals interviewed, an evolutionary pattern was discerned. Often foreign R&D originated as technical support activities for manufacturing or marketing. While not generally considered as R&D, these technical support activities represented a concentration of technically adept individuals whose interests later usually expanded beyond the narrow boundaries of their organizational mission. To the extent that the technical group could gain formal recognition of its expanding interests and capabilities, its official mission often evolved from merely support to one of more responsibility. This frequently occurred through the group's taking on projects not officially assigned to it, until it won acceptance for technical contributions. As a result, a continuum of R&D missions ranged from quality control and technical service, through materials testing, raw materials adaptation, process adaptation, product adaptation, applied R&D, to new product development and exploratory research. One European firm stated that the typical evolutionary process in its laboratories is an initial five to ten years doing quality control and technical service, evolving into modifications of components and intervals, at which level the lab would likely remain indefinitely, evolving later into new product development only if the market required.

Evolution of R&D capabilities and mission occurred in all laboratories discussed but was particularly associated with foreign R&D activities of home-market and host-market firms. Evolution typically represented the only means by which the home-market firms allowed R&D activities to occur abroad—i.e., through the unplanned, gradual aggregation of the skills, capabilities, and missions of technical groups established abroad initially for non-R&D activities. A considerable number of the foreign laboratories of host-market firms evolved also from technical-service missions. As one British executive explained, "Technical service is attached to our foreign manufacturing affiliates and therefore some developmental work reflecting our product lines is always beginning around the manufacturing activities at our foreign plants." Evolution among the foreign laboratories of host-market firms also appeared in several cases in which the laboratories were originally established abroad to satisfy host-government regulations or pressures and were gradually given increased responsibilities in an effort to reduce research redundancy within the firm.

Direct Placement. The term "direct placement" is used in this study to represent the decision to create a foreign R&D group to perform specific scientific and technical missions other than simply supporting local manufacturing and marketing operations. In the case of direct placement, the foreign R&D effort is intended, from its inception, to do R&D. Of the 106 foreign R&D activities of American TNCs, 24 were directly placed abroad. These laboratories belonged to both host- and world-market firms.

Among the host-market firms, the choice of direct placement for establishing foreign R&D activities was motivated by a need to respond quickly to their local markets, by the relative autonomy of their foreign subsidiaries that often leads the affiliate to create the laboratory when it considers it propitious, and by the importance of maintaining good relations with a host government that urged establishment of a lab.

The most severe problem associated with direct placement is one of establishing a good "fit" between the laboratory and the foreign manufacturing/marketing activities. When an R&D group merely evolves, the fit arises out of felt needs. The lack of good fit under direct placement was frequently an attempt to create a full-blown lab, before manufacturing could use its services, or to shift the product line before marketing was ready. Either produced a bad fit—at least temporarily.

World-market companies rely heavily on direct placement for establishing their foreign R&D activities because of their desire to locate in areas where they have access to foreign scientific and technical communities, regardless of their existing commercial activities in that location. An example of the locational influences acting on the placement of research-oriented laboratories was provided by a European firm that has a number of foreign laboratories including four with a heavy research emphasis. In this firm, whereas the *site* and nature of a development laboratory is restricted by the local market demand, the *activities* of each research laboratory are determined by the nature and site of the scientific community from which it can be drawn. This firm's development laboratories tend to follow a predictable evolutionary pattern from quality control and technical service, through minor modifications, and ultimately to some product design responsibility. Development activities are always attached to a specific factory so that they can mesh with the needs of the operating unit. Its research laboratories, on the other hand, are not located near the firm's plants and are the result of corporate rather than divisional initiatives.

Acquisitions. Approximately one quarter of all of the foreign R&D activities belonging to the American TNCs in our sample were the result of acquisitions. All of these belonged to host-market firms. In most cases, the acquisitions were not technically motivated, nor were R&D people involved in the acquisition planning; still, the acquisition was the determinant of foreign R&D activity. As one American R&D executive put it, "Had we not acquired European firms with laboratories, we wouldn't have R&D in Europe." Having once acquired a foreign laboratory, however, the general tendency was to allow it to enjoy its autonomy and continue its work if its products were significantly different from those of the parent firm, or to selectively expand its resources if its projects could complement the existing R&D efforts of the parent firm. This behavior reflected both the lack of R&D's involvement prior to the acquisition and the difficulties of coordinating foreign R&D activities after the acquisition.

Management Style and R&D Abroad

A firm's experience with foreign R&D activities is also distinctly characterized by the way the firm organizes and controls the relationship between its headquarters and foreign R&D groups. The firms in this study had adopted one of four distinct management styles:[2]

Absolute Centralization. Program commitment and total resources used are determined by the parent for the subsidiary.
Participative Centralization. R&D commitment and total re-

sources used are determined by the parent with the subsidiary.

Supervised Freedom. R&D commitment abroad is established by subsidiary's decision; the parent may express its opinion or make suggestions.

Total Freedom. R&D commitment abroad is established by subsidiary alone, with the parent simply being notified or approving pro-forma.

The nature of a firm's management style, for the purpose of this study, was assessed by examining the relative roles of corporate R&D headquarters and the R&D management of the foreign affiliates in matters of resource allocation; program initiation; the selection, monitoring, and termination of projects; and communication and coordination activities. The sample of 50 firms was divided almost equally between some form of centralized and decentralized management styles. In general, the former tended to have rather elaborate reporting, budgeting, and communication systems, and they exhibited a willingness and capability to allocate the corporation's R&D resources on a worldwide basis. The decentralized firms, on the other hand, were considerably more informal in their coordination and control behavior and viewed the R&D activities of their foreign affiliates with almost an air of detachment. European firms were slightly more decentralized than were their American counterparts.

Seven of the firms interviewed could be characterized as being "absolutely centralized" in the management of their R&D activities. All of these firms were American, and only three of them had R&D activities located abroad. ("Absolute centralization" does not appear to be a management style that is easily compatible with the performance of R&D in foreign locations.) Firms in this category tended to be subscribers to the "efficient central R&D facility" school of thought. They tended to separate R&D from the line divisions and to exercise substantial control over the allocation of R&D resources and the flow of information from R&D activities to other parts of the firm. The result is that R&D in these firms was isolated from the rest of the corporation. Communication between R&D and the rest of the organization, when it does occur, occurred informally. In several of these firms there was virtually no travel budget for R&D (particularly researchers), and the principal interactive mechanism between R&D and its corporate constituents ran through the R&D manager himself.

Three of the seven firms exhibiting an "absolute centralization" R&D management style had a home-market orientation. Only one of these three had any foreign R&D activities, which was limited to a minority interest in a joint manufacturing venture that supported R&D activities. Three host-market firms also exhibited an "absolute centralization" R&D management style. They appeared considerably more sensitive to the needs of their foreign markets than did the home-market firms, but only one of them had any foreign R&D activity. One of the world-market firms was categorized as employing "absolute centralization" in the management of its R&D. Although this firm is organized on a regional basis for manufacturing and sales, it has a single R&D group located at corporate headquarters, and a small development group in Canada. The strength of the central laboratory in this firm was revealed by a recent decision to move the Canadian work back to corporate headquarters because it was going particularly well. All product planning in this firm is done for a worldwide marketplace, and products and processes are standardized. This firm favors "absolute centralization" because officials feel that they need "a large nucleus in a laboratory to gain technical synergy and new development, and in this industry to be really successful, you have to start over and develop new products, not modify existing ones."

Of the firms in our sample, including 11 American and 6 European, 17 were characterized by "participative centralization" in the management of their R&D. The "participative centralization" firms tended to exert strong centralized authority over the funding, programs, and often even project-selection decisions faced by their foreign subsidiaries. They also had a large number of overseas R&D activities and employed a structured and often sophisticated coordination and control system. There was, however, some evidence of genuine participation with the foreign affiliates in the management of R&D activities.

The 17 "participative centralization" firms consisted of 15 host-market companies and two world-market companies. Of these firms, 8 were in the pharmaceutical industry while 6 were in chemicals. The relationship between a strong central R&D coordinating mechanism and foreign laboratories with new product responsibility was clearly seen among these firms. A number of them relied on "global product coordinators" who have worldwide product responsibility, including R&D funding and allocation and project-selection decisions, for a particular product line. By using such a focused and formal control system, these firms were able to coordinate closely the R&D activities of their various affiliates yet still encourage the performance of R&D for particularized markets. In support of the highly structured managerial process associated with participative centralization, these firms employed communications systems between R&D and other corporate functions that exhibited a considerable degree of formality. Included among these were: predetermined, structured group meetings and regularly issued formal report summaries.

One of the best examples of participative centralization in R&D management was provided by an American world-market corporation; the two foreign laboratories were funded by the foreign affiliates, but their research programs were established by corporate headquarters. The foreign affiliates participated in determining the pursuit of R&D in this firm through meetings convened by the directors of the laboratories in which they suggested possible research projects and capabilities. The ultimate decision, however, was made at corporate headquarters and was first and foremost concerned with supporting the corporation's total R&D objectives.

Of particular interest to policymakers is the apparent greater willingness of participatively centralized firms to create foreign R&D groups by direct placement rather than by evolution and/or acquisition or joint venture, as this suggests that these firms probably represent the quickest means of establishing full-scale R&D activities abroad.

Twenty-three firms exhibited "supervised freedom" in their R&D management; 14 of these were American, and 9 European. All of these had experience with foreign R&D activities, and half of their foreign R&D activities had missions that included new product research responsibilities. These firms had several overseas R&D laboratories and tended to place primary responsibility for operational decisions in the hands of the foreign R&D management. Coordination appeared far less formal among these firms than that attempted by the participatively centralized firms and relied on good interpersonal relationships among laboratory directors and much travel. The corporate R&D staffs at the headquarters of the supervised-freedom firms were typically responsible for reviewing the intentions of the foreign R&D groups and for assuring corporate management that the foreign groups were performing at an acceptable level.

Two of the supervised-freedom firms had home-market orientations. In one case the considerable freedom vested in the foreign laboratory was a result of the small size of the laboratory; in the other, it resulted from the wide diversity of technical interests of the firm and the fact that it established its foreign laboratories 50 to 20 years ago as independent entities; this tradition of independence has remained.

Of the supervised-freedom companies, 19 had host-market orientations, 9 of these being American. Their coordination is illustrated by a European tobacco firm whose overseas affiliates are profit centers that perform whatever research they need. The corporate R&D group, in this firm alone, provided a professional critique, although it did review programs from a fiscal and operating standpoint. There was no formal corporate research plan, and the foreign affiliates enjoyed considerable independence. A defense of supervised freedom in the management of R&D was offered by an R&D manager in a European food company who observed that "R&D needs to be a self-starting activity, which cannot be built up or cut back quickly through excessive budget changes." Presumably, the supervised-freedom firms believe that such stability is worth gaining at the expense of being able to coordinate the corporation's R&D efforts on a firm-wide scale.

The firms exhibiting a supervised-freedom R&D management style tended to have central R&D groups that played an important role in assuring top management of the practicality, utility, and soundness of the R&D being performed by the subsidiaries. As one R&D executive stated: "I serve as a radar antenna for top management, to make sure our technical programs make sense and are done right. I review plans, projects, expenses, and personnel movement. I have dotted-line authority over each technical director." But within this firm the actual R&D project and program decisions were made at the division, not at the corporate, level.

The importance of central R&D laboratories to supervised-freedom firms arises from their performance of long-range research that would not typically be pursued by the operating divisions. In one machine-tool company, corporate R&D focused entirely on new ventures rather than extending existing product lines; in a chemical-based firm, central R&D worked only on projects of a common interest to the operating divisions, yet which none of the operating divisions would pursue independently.

In terms of communications and coordination, the supervised-freedom firms relied on informal relationships, with far less use of routine formal meetings than found in participatively centralized firms. The supervised-freedom firms did, however, sponsor considerable travel by individuals on an ad hoc basis for coordination purposes. The reliance that "supervised-freedom" firms placed on their central R&D labs, combined with the heterogeneity existing among their affiliate labs, as a result of their relative freedom, could, however, lead to selective restriction of scientific and technical communications within these firms. Communication networks that were indirect, by virtue of having to pass through a central laboratory, were shown to offer an opportunity for selective interference, whether intentionally or unintentionally.

Two of the firms exhibiting supervised-freedom management styles had world-market orientations. Both of these firms are large, American, electronics-related companies. In both cases, their foreign R&D laboratories were performing highly sophisticated, fundamental research and were relatively self-sufficient within a specific technical area. The foreign laboratories were all originally established to gain access to existing foreign scientific and technical communities. The combination of technical self-sufficiency and the collegiality found among research laboratories in the world-market firms had led to the adoption of a supervised-freedom R&D management style.

Only three firms among those interviewed could be characterized as exhibiting "total freedom" in their management of foreign R&D. All three firms had established R&D abroad through acquisition, and in two cases the skills and interests of the foreign laboratories were considerably different from that of the central corporate laboratory. All three firms in this group recognized the costs associated with their present method of managing R&D and its resulting inefficiency. Our interviews suggest that total freedom in the management of R&D is most likely the result of a transitional situation and is not a consciously adopted policy.

Science Level and Managerial Style

Market orientation, while useful for understanding R&D location decisions, is not a good predictor of management style. Industrial scientific orientation, on the other hand, did appear to influence management style significantly. Firms in science-based industries such as electronics, pharmaceuticals, and chemicals favored a higher degree of centralized control over R&D within the corporation (yet with some room for participation by the operating divisions) than did firms in industries that were less science based. This difference most likely reflected the importance of accepted international standards in the science-based industries as well as the numerous products they sell and the sophistication of the R&D in these industries. As one executive observed, "If your research is related to products and you have myriad products and markets, you need a high degree of coordination. When you're dealing with one or two metals, you're probably familiar with everything going on, and you can enjoy more informal coordinating schemes."

Factors Affecting "Critical Mass" in Foreign R&D Groups

While the size of any particular R&D group is a function of many variables, there is substantial agreement within the R&D community that there exists a "critical mass" of R&D professionals for any laboratory that must be reached if the laboratory is to be a worthwhile investment. Critical mass is that size necessary to ensure rich communications both within the group and between the group and its environment, to provide the scientific and technical capabilities among the personnel necessary for fulfillment of its mission, and to acquire whatever instrumentation and organizational slack is necessary for acceptable performance. While there is considerable variation in the estimates of critical mass for specific situations, several insights regarding critical mass developed in the interviews.

R&D laboratories in industries serving consumer markets required a smaller R&D staff to reach "critical mass" than did laboratories in science-based industries. R&D groups in consumer-oriented industries required less sophisticated personnel and less variety in personnel specialization than did R&D groups in science-based industries. R&D on manufacturing process required smaller R&D groups, less technical specialization, and less contact with external organizations than did R&D on new products. All these results have particular relevance to the desirability and likelihood of TNCs' establishing R&D laboratories in foreign locations, and suggest that consumer-oriented and process-related R&D are the forms of R&D that TNCs are most likely to place in developing countries if and when they make such a location decision.

Product and Process Standardization and "Appropriate" Technology

Standardization of production processes and products appeared to be a fairly common objective among most of the firms interviewed. While there was some willingness to modify a production system so as to build in more labor-intensive activities at the periphery (e.g., materials-handling systems), no firm appeared willing to make a major change in its production processes to accommodate different foreign resource factor endowments. To the extent that significant adaptations to process machinery or product design were made, they were typically a function of market size rather than local resource endowments. Similarly, almost all firms preferred products that could be offered on a worldwide, or at least regional, basis. Even host-market companies indicated a growing inclination to discourage products unique to a particular rational market unless the market was of considerable size.

Collaborative International R&D Arrangements

Collaborative R&D arrangements can offer TNCs an alternative means of pursuing R&D abroad instead of direct placement, evolution, or acquisition of an existing foreign laboratory. Considerable use of collaborative arrangements leading to R&D activities was found among the firms in our sample. Most frequently, this took the form of joint manufacturing ventures between organizations of differing nationalities, with R&D representing a function of this joint venture and performed by either the joint venture itself or by the parent company labs in cooperation. The key determinant of *active* R&D participation by the companies in such an arrangement was the competitive advantage to be gained. Foreign joint ventures generally gave rise to *new* R&D only when the firm could maintain control over that R&D; namely, when it enjoyed a majority interest in the joint venture. Minority ownership positions and/or the collaborative presence of a competitor posed sufficient risk to proprietary R&D interests to restrain the firm from active participation in R&D labs.

In those cases where the TNC did not actively support the joint venture's R&D efforts, the level of their technical involvement ranged from making existing technology openly available to the joint venture to treating the joint venture as if it were any other potential customer. Slightly less than one half of the joint ventures identified in the study represented full-fledged commitments by the parent corporations to participate in supporting the joint venture's R&D efforts.

Several instances of joint R&D ventures, occurring outside a manufacturing-related context, were found during the study. Such arrangements took the form of cooperative research activities at one end of the spectrum and sharing of R&D plans and information of the other. Among the former were both multilateral and bilateral project-oriented R&D ventures while the latter was typified by technical exchange agreements. The impression gained from the firms interviewed was that although it is unusual for R&D of a really substantive nature to be made available through such joint R&D ventures, many firms are involved in them because the cost to the firms in terms of potential loss of competitive advantage is low.

In a few instances, the company felt it needed the expertise of the partner, and in others, it was a means of cost sharing.

The Diffusion of R&D Capabilities to Local Scientific and Technical Communities

Considerable evidence was found of the diffusion of R&D capabilities and skills from TNCs pursuing R&D abroad to host-country scientific and technical communities. The extent of this diffusion, however, was typically not large, nor was it the result of conscious decisions by the firm.

Almost all the firms interviewed employed nearly 100 percent host-country nationals as professionals on their foreign R&D staffs, and many of the firms had local nationals serving in first-level R&D administrative positions. Some training of personnel did take place, though typically not through formal programs. Because of the initially high educational level of R&D professionals, the training added little to their theoretical, but substantially to application, skills. Turnover among foreign R&D professionals was quite low, and when a professional left the employment of a firm, he/she typically took a position with another TNC rather than going back into the local business community, government, or universities.

Evidence of technical upgrading of local suppliers and customers was also found. Examples of such activities include: active participation in market development, training programs, technical assistance missions, improving local quality-control skills, location for development of local sources of raw materials, and assistance in improving local warehousing and distribution systems. Typically this occurred when it was necessary for smoothly integrated operations for the transnational corporation. Support for local university activities and assistance to host-country government agencies were also reported, but there was no evidence that such activities were frequent or significant. The impression obtained was that assistance efforts occurred due to chance and/or confluence of interests rather than by plan.

Host-country Government Policies on R&D Location and Performance

To focus on government policies toward R&D by international companies, we asked each firm the questions:

1. Has your firm received pressure from foreign governments to locate R&D abroad?

2. Are foreign government inducements/pressures successful in attracting transnational corporate R&D activities?

The standard response to each was no, but the results of our interviews indicate quite strongly that the situation was actually otherwise. This contradiction results from a comparison of the respondents' ranking of factors in discussion of the motivating factors that influence the R&D location decision—these indicated sizable host-government influence in their decisionmaking— contrasted with an almost blanket negative to a point-blank question on the issue.

While host-government interference was not as important as market potential in determining foreign R&D locations, it probably was more important than the condition of the local scientific and technical infrastructure. Furthermore, government inducements/ pressures to locate R&D appeared to be quite effective in attracting foreign R&D presence in developed and advanced-developing countries.

The interviews for this study indicated that countries with attractive markets could compel firms to perform some R&D locally. Unless, however, the country had an adequate scientific and technical infrastructure and the promise of sufficient market volume, the performance of any R&D in that country by a transnational would be low level and just sufficient to meet government regulations. Furthermore, it would appear that if government regulations increased without an accompanying improvement in the local scientific and technical resources available to support additional work, the TNC would seriously consider exiting from the market.

Our interviews also indicated that the developed countries have tended to rely on positive inducements such as tax subsidies, while the developing countries have more typically employed negative inducements such as import restrictions and profit-repatriation constraints. The behavior of the developing countries may well reflect their past unsuccessful experiences at trying to attract the R&D activities of transnational corporations. The negative inducements relied on, however, are best suited for those instances in which the target firms have already established a local operation of some sort and whose mobility, therefore, is somewhat limited. In practice, these inducements have only moderate efficacy in influencing firms already located in a developing country to do R&D there and serve to discourage potential investors who have not yet committed themselves.

Although the developed countries tended to employ more positive inducements to influence R&D than was found in the developing countries, these inducements were also of doubtful efficacy and did not appear to figure prominently in the R&D location decisions of any of the firms interviewed.

SUMMARIES OF COMPANY CASES

The five case studies were the initial step in the overall study and formed its backbone in the sense that experience of these companies and the problems that they face were used as the foundation for the interviews with numerous other companies and initially with government officials as well. In each instance, the headquarters research unit was visited—whether in the form of a central research lab or merely a headquarters coordinating unit— plus at least one company lab in an advanced country and one in a developing country to be able to compare the gowth paths and the different methods of coordination and management control between the two and thus be able to draw lessons as to future patterns likely in LDCs.

Each company visited went through substantial R&D reorganization after the interviews were conducted, and serious questions arose as to whether or not the cases should be updated before publication. Since the updated cases would probably be out of date by the time they were published, the decision was made not to change the original data since the information was accurate as of the time of the interviews. The cases are written in the present tense (as of the time of the interviews) but are in fact a snapshot of the way in which R&D activities *were* conducted and coordinated during 1977 and 1978. They are not representative of present structures or procedures and should therefore be considered as illustrative rather than as presently descriptive. Since it was the purpose of the research to obtain real illustrations of overseas R&D activities, the cases served their purpose well, and they can still be considered as illustrative of the types of problems faced and practices employed. For this reason they are included in the final study, and significant lessons have been and can be drawn from them. Only three of the five are included in the final study since publishing all five would have lengthened the study excessively and the lessons were amply illustrated in the ones chosen for publication.

Du Pont's Department of Fabrics and Finishes

Du Pont is organized by product division; the "line" organization is comprised of a number of "industrial departments" and an international department, with the latter being liaison between the industrial departments and some of international activities. In addition, there are several staff departments, three of which are of particular relevance to R&D activities: an Information Systems Department, an Engineering Department, and a Central Research and Development Department. Central Research conducts a wide range of basic research in the United States and maintains the administrative structure to house departmental research on a "research campus" at the Experiment Station where common support facilities (such as a library and sophisticated equipment) can be used by all departments.

The Central Research and Development Department has nearly 700 professionals, of which 500 are technicians and 200 are Ph.D's. The Ph.D's. are predominately in chemistry, though there are a number of physicists and a few chemical engineers. Results of the Department's work are written in research reports, which in turn are sent to the relevant industrial departments of the company and to various affiliates overseas on a "need-to-know" basis.

The industrial department chosen in consultation with the company as providing the basis for a useful study was the Fabrics and Finishes Department (F&F), which comprises a wide range of paints, plastic coating, adhesives, waxes, industrial finishes, and consumer products. Its business, which concerns auto finishes, refinishes, and industrial and trade paints, is supported by a substantial R&D activity headed within an R&D division of the department. This division serves all the activities of the Department out of its five different units: a lab at the Experimental Station in Wilmington doing basic research; another at Troy, Michigan, specializing in auto finishes and refinishes; the Marshall Lab in Philadelphia, concentrating on trade and industrial finishes; various units working on production technology; and a patents and licenses unit in Wilmington. The R&D activities of the F&F Department were not brought together into a single lab until 15 years after the establishment of the Department, and the present R&D division was set up only in 1970, with its program essentially determined by the marketing division of the Department. By 1974, there was a gradual shift to a separate divisional program determined by the R&D director in F&F, looking toward longer range objectives and the health of the business as a whole rather than emphasis on marketing.

The various units of the R&D division develop products and processes and support their adoption overseas; when adoption occurs is determined largely by decision of the affiliate abroad, but in consultation with F&F in Wilmington. The market orientation is toward the host country or region. Foreign labs will receive the appropriate reports of U.S. labs and will proceed to make any modifications necessary. The scope of an R&D project in paints begins with the determination of qualities of a paint needed—durability, resistance to cracking, color-fastness, drying time, flexibility, resistance to heat and cold, and so on. Once the paint is developed, research is still required as to the best process of application of the finish to the surfaces to be painted. A third stage relates to pollution controls both in the processes and in the use of the paints. The labs also undertake projects on the analysis and evaluation of paints by competitors around the world.

To facilitate adoption of appropriate products and processes by foreign affiliates, and to assist in the processes of adoption and adaptation, personnel from the central labs make frequent visits abroad. Reflecting an approach of "supervised freedom," there is a special liaison officer in the international operations part of F&F who makes certain that communications are open and effective, requiring numerous visits on his part to the various overseas operations. In addition to these visits, there is a biannual conference of international technical managers, attended by one or two representatives of the R&D labs of foreign affiliates, meeting with professionals from the U.S. labs of F&F Department. In addition, longer term exchanges of personnel occur both for training and for providing needed technical or managerial assistance.

The affiliate in Belgium was begun in 1959, with only quality-control and technical-service facilities. However, the marketing group soon recognized that European customers would not readily accept the U.S. paint line; worldwide standards are not feasible in paints, and product adaptations were required for local uses and materials. A lab was opened in 1962 for development of finishes, primarily oriented to color matching and customer technical service. By 1966 the lab had doubled in size and moved into the developmental stage in paints, and two years later it began applied research for new product adaptations. By 1977 the lab was four times its original size, encompassing substantial new facilities but remaining at the level of applied research. The growth of the lab has responded to the pull of market demand in an evolutionary process. Its expansion has tracked fairly closely with the growth of market sales; its budget has remained 3 percent of European-wide sales by F&F.

The geographic location of the lab was easily enough decided on—

it was simply put next to the existing manufacturing plant. The location of the manufacturing plant was determined by the market—mainly General Motors' affiliate in Belgium. The projects undertaken at the lab reflect the specific market demands and the ability of the company to support R&D costs out of its own revenues. Substantial technical service for customers is required because of the fact that each batch of paint varies even with identical inputs and with continued efforts to achieve standardization of materials supplied to each customer. The lab contains a variety of test facilities and equipment permitting formulation of paints and testing to meet specifications required by customers and governmental standards.

Personnel at the lab are encouraged to advance their careers through taking advantage of educational opportunities. The company pays the costs of such education and the salary of the worker during his course work. Not many people take advantage of this opportunity, essentially because there is little mobility either into or out of the lab, given the governmental restrictions of laying off personnel. These restrictions make it difficult to let anyone go, and consequently there are few openings unless there is an expansion of the lab's program.

The scope of the lab's program is defined as "developing projects for the needs of customers," which the lab is capable of doing in the main. When it cannot, it turns to the U.S. labs for help. European customers use different paints and application procedures, making "in-line" repair difficult. The application procedure is hard on the paint itself, which must be durable enough to stand the punishment.

Because of a European preference for enamels, the lab in Belgium had to go into alkyd-enamels to service the needs of the large dealer repair shops, plus independent and small garages. To get into the European market, the lab began with a comparative analysis of European paints. The Wilmington labs sent an experienced polymer chemist to work with a Belgium chemist, plus a Belgium resin chemist and two technicians. They worked for six months to develop a satisfactory polymer for European needs. Eighteen months more were required to develop colors for the basic color line; and the polymer had to be reevaluated in each of the color ranges. The lab then began field testing and ran into a dulling problem caused by "foul gas" from shop heating systems. The lab had to modify the formula so that it could perform in adverse conditions.

European demand for high-quality paints is so great that customers will often pay a proportionately larger premium for paints if they have a 10-percent increase in their properties such as hardness, flexibility, glass, durability, appearance, adhesion, resistance to chemicals, and nontoxicity. It is of course impossible to achieve all desired levels of these qualities in a single paint since to increase one may decrease another. To determine the trade-offs to make and therefore how to guide the research projects, marketing managers constantly research present and future demand. From these data, R&D objectives are formulated jointly by marketing managers and research supervisors. The R&D lab does not have to accept all marketing requests; when it does agree, it sets a reasonable schedule and is expected ·to adhere to it in meeting the needs of the marketing group.

Most of the formulations in the Belgium lab are based on those generated in the United States, but modifications are frequently required to permit substitution of local materials, which have different qualities and reliabilities. The substitution of local materials requires a breaking down of the U.S. product into its components and a substitution of local inputs as close as possible, though on a trial basis, leading to numerous trials to determine suitability.

The translation of U.S. formulations and procedures is facilitated by a weekly flow of reports and staff notes from the United States to Belgium. And monthly reviews from all F&F lab locations from around the world provide important insights into what others are doing. In one instance, the Belgium affiliate picked up an auto paint line that was developed by the U.S. lab but which F&F had not commercialized there. The Belgium plant adoption arose from visits of chemists among the labs. The Belgium professionals considered that the new paint would meet customer requirements in Europe, even if it did not in the United States. At times the Belgium lab has had to develop new products, based on U.S. technologies or processes, to meet markets that are different from those in the United States.

A Brazilian affiliate, Polidura, was acquired in 1972 by Du Pont. Polidura had just begun to establish a lab and organize an R&D division. (The company was family owned, and the founder was approaching retirement.) A quality-control function already existed in the manufacturing plants; also a small lab for raw materials testing. At the time of the affiliate purchase, there were 20 people in the lab, 3 of whom held degrees. By agreement, Brazilian management remained in charge of the company, and from 1972 to 1975 the lab activities grew little since the market for paints boomed and would take anything that anyone could produce.

After 1975, as the market declined cyclically, Du Pont took over the management and brought in not only new products and process technology but also its R&D management and professional orientation. The research activities were reorganized to bring them into line with Du Pont standards of operation and F&F's scope of activity. The R&D lab now comprises research groups in resins and polymers, one in primers and wood finishes, one in water-base/electrocoating, and trade paints, and one in automotive top coats. A dispersion lab is concerned with process engineering; a plant-support lab is responsible for quality control and for making adjustments in the batch processes; a materials lab analyzes inputs; an automotive lab concentrates on customer needs in those paints; and a marketing-support lab provides technical service for trouble-shooting with specific customers.

This separation of functions had as an objective the determination of specific costs of research and technical support, which had been all lumped together with manufacturing costs by the prior owner.

By 1978 the staffing of the labs had reached the critical mass, amounting to 160 professionals and auxiliary personnel, with 36 of these holding degrees. The largest portion of the auxiliary personnel are in quality control and sales support.

To help train the professionals, several of them have been sent on different occasions to Du Pont labs, and a few have taken courses in São Paulo, offered by the Association of Paint Manufacturers. Much of the training has to be "on the job," however, because universities are not oriented toward industrial research.

The mission of the lab is to develop products out of the Du Pont line that are new to the Brazilian affiliate. This task requires an identification of the products that are appropriate, the substitution of local raw materials, and changes in formulations needed to meet customer requirements. A process engineering section will be formalized later. And the lab has had to reorganize and restructure management responsibilities since it has grown too large for everyone to report to the company manager as was the practice before it was purchased by Du Pont. It required nearly two years to form the present organization of the lab, and the lab director spent a month in Wilmington becoming familiar with company procedures. Although procedures at the lab are not yet set up the way they should be, new methods are being introduced concerning administration, relationships to manufacturing and marketing, and program control and coordination. Adoption of these procedures will facilitate more effective communication between the affiliate and parent organization and the United States. Formal courses on analysis and procedures have been provided each new entrant into the company, lasting for a three-month period on materials, intermediates, administration, and organization. Another three months of training on the job is required before a new professional can be fully useful.

To facilitate the flow of communication between lab and the parent organization, three channels are open: the technical manager of international operations of F&F who provides continuing liaison; a contact through the international department; and a flow of technical reports and formulae, which reduce trial and error in the Polidura lab.

In addition to assistance from the United States, Polidura had an exchange of technology with the Belgium lab; for example, the Belgium lab sent a technician to explain a heat-drying process, which the Brazilian lab later copied, saving considerable development effort. Also a new line of finishes will rely on the Belgium line since it is similar to European paints, which are used by several manufacturers in Brazil.

In its early years, the materials section of the lab was dedicated 60 percent to quality-control work, but it will later expand its activities in the improvement and substitution of local materials, and the quality-control work will probably be shifted to the manufacturing plant and out of the lab itself. This will permit an expansion of activities without an increase of personnel in the lab. Further changes in the program can be anticipated as the market changes, and eventually the lab will be reoriented more toward the various marketing demands, responding with new formulations and processes. Du Pont's future growth in Brazil will be evolutionary, responding strictly to market developments and shifts, and the growth of the lab will track with this process, as in the Belgium lab.

Johnson & Johnson Health-Care Products

We selected the J&J health-care product line for study since it is the major business of the company, and foreign labs existed in two advanced countries and one developing country. The basic managerial style of the company is one of decentralization and virtually complete independence among the various company units; there is no central research lab for the company as a whole. Each of the separate companies have set up whatever R&D facilities it deemed appropriate, and each of the foreign affiliates is basically free to determine the nature and extent of its R&D activites related to its business objectives. No mechanism exists for coordinating research activities worldwide, and each company is essentially a self-contained unit, though each is expected and is in fact willing to pass along any information desired by any other affiliate.

U.S. Labs. Labs within the United States are essentially divisional labs, reporting to their operating divisions directly, though within the Domestic Operating Company (DOC). There is an Exploratory Lab doing some applied research for all of the divisions. A vice president for R&D of the DOC is a member of the management board of the DOC, and R&D directors of each product division report to him for coordination.

The divisional labs are responsible for developmental research in their product areas; the Exploratory Lab does applied research aimed at developing specific products or solving specific problems that are not necessarily part of any given division. Basic research is simply not done within the DOC labs. There are about 100 personnel in the Exploratory Lab, encompassing professionals from biology, microbiology, veterinary medicine, bioengineering, physiology, biochemistry, organic chemistry, histology, pharmacology, pharmacy, polymer chemistry, and rheology. About half are Ph.D's., and the rest have undergraduate degrees. The funding for exploratory research comes from an allocation by the four operating companies in health-care products. These contributions usually run less than 1 percent of sales.

The program of the exploratory unit is over half in product development activities aimed at discovering new business opportunities inside or outside of the present product lines of various divisions—within the health-care area—and the other half in gathering scientific information to provide innovative leads or product development in the future. Frequently the divisions are reluctant to pick up products coming out of the exploratory unit; only if there is competitive pressure in the market do they become more eager. This product/market orientation is reflected throughout the entire corporation, with R&D program control being the responsibility of each division or company.

Since overseas labs are independent of any control from the United States, communication from any of the U.S. labs is through a formal exchange of information or personal visits to see what might be of interest. The value of the outward communication from the U.S. labs is evidenced in the fact that virtually all overseas products in the health-care field have stemmed from the DOC output. The establishment of personal contacts is stimulated by meetings of international officials, permitting them to exchange ideas and to become more readily acquainted with each other and therefore to communicate more easily by telephone and telex. Further, to facilitate this communication, any newly hired overseas R&D director is brought to the United States for training and background information on the company's product line and technology.

The Health Care area has a lab, the director of which reports to the divisional general manager, who is in turn responsible for brand management. However, during the course of the writing of the case, this reporting channel was changed so that the R&D directors report directly to the vice president for R&D of the DOC itself.

Given the independence of the labs, ideas generally tend to be initiated from within, and any new proposal must be approved by the director of R&D, the general manager of the division, and the brand manager of particular products. A "critical-path book" is kept on each project; it reports on a quarterly basis the accomplishments to date and the objectives sought for the next three months. When finished, the project produces a "fact book" that can be used by anyone in the company to develop the product. It also includes specifications and must be followed in marketing and advertising the product so as to stay within acceptable claims.

The lab is supported by the local science community, relying to some extent on the university labs at Rutgers and Princeton, as well as scientific developments coming out of MIT. Its professionals are also members of various associations, which keep them informed of the latest developments.

The Health Care lab does not provide *continuing* support for the overseas labs, though it has at times been asked for trouble-shooting assistance. In the main, communication occurs through visitors who bring everyone up to date on successes and failures.

The patient-care division includes products related to the care of patients in hospitals. It covers not only the developmental research on these products but also clinical research and technical support for sales and marketing. Of some 47 personnel, 31 are professionals from a number of chemical fields. The expertise of these individuals is drawn on at times in support of programs in the division's overseas labs. Also it retains over 30 consultants on various assignments during the year, extending from hospital needs to space science. This expertise could not be obtained on a permanent basis within the lab, and without such assistance from the science community, several of the projects could simply not be carried out.

Differences in approaches and product lines overseas (lack of worldwide standardization) have prevented significant joint projects in the United States and abroad among development labs within the DOC. Recently, however, the tendency toward further decentralization within the R&D division has been reversed, reintroducing greater coordination into what had become a fairly competitive system overall. It appears that the overseas labs will, however, remain relatively isolated from their U.S. counterparts, mainly because of geographical distance and infrequent personal contacts. Given this distance, marketing divisions overseas are slow to pick up products from the United States, taking a conservative view of the use of their funds and time. Marketers tend to be more interested in nuances with existing products than with completely new items, which are unfamiliar and risky. Consequently, such new products have to be pushed either by top management or from other interested areas of the company. Since the main source of new products for any of the foreign affiliates of J&J is from among those marketed in the United States first, the major problem is that of communicating to the affiliates what is available and how useful it might be in their own markets. Meeting this problem requires identifying market needs, which is not always adequately done by the overseas affiliates. The company faces the problem of whether to establish a central technical group at headquarters to produce prototypes and finish products for overseas affiliates, without having itself a close tie to the foreign market. Alternatively, a smaller technical group could be tied to one of the existing overseas affiliates—closer to the market, but possibly without all the scientific expertise or backup needed.

British Lab. The British affiliate of J&J was formed in 1924; thirty years passed before establishment of an elementary lab in 1954; a central lab for the company was established a few years later. This lab was initially directed by a professional, oriented toward academic-type basic research, and he brought in a half dozen Ph.D's., who remained remote from manufacturing and marketing. Consequently, the lab was closed in the early 1960s, with the principal scientists' departing for positions in other companies. Since it takes at least five years to get any new products out through the R&D process, J&J officials assert that a lab must be closely integrated with the rest of the company objectives, or else top management will consider it a "black art" with immeasurable, and possibly undesirable if not useless, results.

In the 1960s a new lab at the British affiliate evolved out of the quality-control and technical-service operation at the plants, responding to local market growth and needs. Scientists had to develop an expertise and knowledge of how to modify products and develop those suitable to the local market, so some of the scientists in technical service were spun off of that operation into a new lab activity. The personnel in the lab have remained essentially British, working with British companies as customers, and this independence permits the lab to recruit good researchers on the basis that they will be doing something for a British company and its British customers—rather than working for a U.S. parent company for worldwide customers.

The British company has four manufacturing units under it, not all of which have R&D capabilities. There is a company director of research, having responsibility for quality control and technical service, plus the departments of microbiology, product development, and medical assessment.

Critical mass was reached with some 25 professionals who cover five major areas of research and are supported by some 20 technicians and a number of other administrative and support personnel. Just under half of the professionals have Ph.D. degrees in several scientific areas. There is a surfeit of Ph.D's. in Britain, but they are not trained adequately in industrially related research. Upon entering the company, the new researchers are given a short orientation course and then put to work on the bench. There are no inhouse courses by which to educate them further, so they must learn on the job. However, scientists are encouraged to take courses in local schools or universities to upgrade their skills.

The budget for product development in the British company is not determined by any rule of thumb compared to sales. Rather, the budget is determined by a set of proposed projects, paid out of corporate funds, which are then costed back to the several marketing divisions pro rata. Each project will be checked first with the appropriate marketing group, and the R&D division will try to sell it on the usefulness of the project, and the marketing group must make a decision to accept it or not. In some cases, the R&D division may proceed anyway if it thinks the project is valuable; if marketing does not eventually take it up, the technology or product may be licensed to another company. Lab officials consider that if they stifle such initiatives on the part of researchers, the latter will not keep interested and on their toes.

The product line of the British company is essentially that of the United States; but it has to make some modifications and has at times adopted products not in the U.S. line. In one case it picked up a "failure" in the United States that was suitable to the British market. In another case, the materials necessary to make a particular idea work were available in the United Kingdom but not in the

United States, so the U.K. company picked up the idea and pursued it to fruition. From these marketing and resource differences, J&J managers have drawn the lesson that local R&D activities should start from market-oriented needs and respond independently from the activities elsewhere in the company. Only if markets in *all* countries were basically similar would it make sense to allocate to affiliates pieces of research projects that were directed and controlled from the center.

Given this independence, there is a tendency among labs in foreign affiliates to take on more than they can complete. They then must make a trade-off between reducing the number of projects or possibly reducing the chances of success in a larger number of projects. The trade-ofs are eased by transfers of professionals among various labs for the purpose of working on what others might be doing so as to bring those developments back to the affiliate. For example, the British lab had a man in Canada learning about nonwoven absorbent technology for use in a new cloth essentially for bags. Despite the relative independence of the labs, the managers in the U.K. lab consider that it would be extremely difficult for them to operate without reliance on the labs in the United States, which provided not only a variety of new products but a number of experiences on continued changes in the existing products. Given this broad reliance, managers found it impossible to calculate the value of the contribution that they received from the center.

As with the U.S. labs, J&J officials in Britain observed that it was quite necessary to have a scientific community within the host country to provide support not only to the scientific approaches on the part of the lab professionals but also to provide governmental regulatory agencies with impartial scientific advice and counsel. The company also taps this community for consultants, spending £ 22,000 per year.

Canadian Lab. J&J's affiliate in Canada was, until the early 1950s, simply a manufacturing extension of U.S. operations, copying in detail what was done in the United States—though with considerable lag. No scientific effort was mounted to organize the transfers of technology, much less of R&D results.

However, in 1953 the president of the U.S. company ordered a lab to be established in Canada, despite the fact that the Canadian management did not see the need for it and had in fact procrastinated over its establishment. The president considered that "not all the brains are in the U.S.A." and saw the future need to use local raw materials, adapting them for production in Canada. The new director of the lab hired several Ph.D's., but they were not oriented toward industrial research and promptly had to go into the plant to find out what it was making and how it was done. Having no lab facility, they were physically located in the plant, which enabled them to see how production could be improved and costs minimized.

Consequently, quality assurance and technical service were placed under the lab director, and this remained the first charge of the lab for a while. Its first task was an inventory of plant products, procedures, and marketing approaches, requiring three to four years just to find out what standards were being used in production in the uses of the products and the markets. The lab professionals spent the first years in writing up specifications of what was done and how it should be changed to accord with higher U.S. standards. The introduction of these standards was not easy, and the head of quality assurance was nearly fired for refusing to approve a shipment of repellent paper that he considered deficient. Management had to be sold on the desirability of strict quality assurance. But it was needed, for at the time the company produced four or five different absorbent cottons without it being clear what the different production standards or uses were.

These first ten years were a period of frustration for the R&D personnel since they could not conduct research as long as they were embroiled in the day-to-day operations of the plant. But this lab capability was needed to enforce quality assurance in the company and gradually raise its interest in R&D efforts.

The lab began new product translation and adaptation only in 1958–59. It drew the original products from the United States, bringing back sufficient information so that the Canadian management could see the value of the new products. Since the production of fibrous products for industrial clients is by far the largest product line of the Canadian affiliate, a lab was set up with units concerned with technology developments, product development, and operation development. In addition to health care, a lab was set up with units responsible for product development, operations development, and evaluation of competing products. (The organizational structure of such a lab depends on its mission, relationship to the rest of the company, and the personnel involved.)

J&J officials stressed that R&D personnel are not born, they have to be made. Not only must they be made in the sense of learning how to do industrial research, but they also must be fitted into an existing organization, which has its own goals and procedures. Since the company is located in Montreal, the personnel of the lab come from universities in the province of Quebec plus some from other labs and other industry sectors, and some from labs of other U.S. affiliates that have been closing down because of poor R&D management. A six months' apprenticeship is provided for each university graduate coming into the lab so as to provide an appropriate orientation. Support is also given for continuing work toward formal degrees, and there are some inhouse programs on management, communication, and planning skills.

No special method exists for determining the budget of the lab. The budget is essentially based on the usefulness of individual projects. Total R&D expenses, as a percent of sales, have dropped in the past few years simply because sales have gone up significantly. The overall budget has increased by over 80 percent during the years 1972–1976, with total personnel rising from 80 to nearly 100, of which half have been in quality assurance and the rest in R&D activities. The lab cannot grow much more rapidly simply because of the difficulties of assimilation of professionals, despite the pressure from top management to expand quickly. Although the lab will continue to grow, it has already reached the critical mass necessary to permit it to do applied research. However, to reach this level required over twenty years of evolutionary growth.

The lab has begun some new product development, finding a gap in the U.S. programs that it could fill. As an example, the Canadian affiliate sought a "poor man's" technique for producing a nonwoven fabric useful for the limited market in Canada. It needed a new product with different manufacturing processes, applicable to small-scale operations. The U.S. company had produced a fabric by methods too expensive for Canada, and the latter needed a product that could be produced on existing equipment so it would cost no more than $250,000 to tool up. The project was to meet also the requirement that the new product should fit at least two market uses in the existing J&J line so as not to have to develop a completely new market. Through the ingenuity of some of the researchers, it was able to produce a new product through an air-blowing technique experimentally employing a redesigned vacuum cleaner for lab equipment.

Although the Canadian lab is completely separate from the activities in the United States, it has been able to get anything from the United States that it knew how to ask for. Its officials have been treated as colleagues rather than second-class citizens when they visited the United States, and a continuing and productive rapport has developed. The lab simply could not have gotten where it has without assistance from the U.S.; nor could not have found similar assistance within Canada since such knowledge simply did not exist there.

This support at times turns out to be mutual, with new developments flowing back into the United States. The Canadian lab had been working on nonwovens with an air-filter equipment company. Out of many experiments came a new machine for the manufacture of unique nonwovens for diapers and for other products. The lab failed to persuade manufacturing that it would be useful technique for Canada. It was, however, picked up by a J&J company in Chicago, and one of its engineers spent six months in Canada working on the project, eventually taking all data back to Chicago. In another project, research personnel from Europe have come to Canada to learn what was being done. Lab officials recognize that being so close to the United States is a potential obstacle to the development of the lab because of potential dependence on U.S. work, but this disadvantage is offset by having full knowledge of what is going on in the States and being able to work from those results.

One of the complications of the closeness between the United States and Canadian operations is that the marketing divisions in Canada are themselves very strongly tied to the marketing divisions in the United States, taking their signals from them. Canadian marketing, therefore, tends to reduce its risks by following the U.S. pattern, rather than taking initiatives that might come out of the Canadian lab.

Finally, the Canadian lab does not have the same support from the science community in Canada that the labs in the United States and Britain have. This is because there has been little academic interest in industrial research. It does have a few consultants, however, but only two are on retainer. In the view of lab managers, the development of a local science community would take some burden off the lab itself.

Brazilian R&D Activities. The Consumer Division of the J&J affiliate in Brazil produces some 60 products in the baby product line and household health and cleanliness. The company was begun in 1943, but it was only after World War II that quality-control activities were instituted and in 1955 that a director of research was appointed. He had both product-adaptation and quality-control responsibilities, requiring some 30 people in the various technical activities by 1960. By 1972 the lab reached its present organization, and a new laboratory facility was provided.

The lab activities are now divided between consumer and hospital products on the one hand and medical and diagnostic products on the other. (Other divisions produce other products in the J&J line in Brazil.)

Qualifications of the staff and the lab have increased, moving from technicians to professionals with at least bachelor degrees. There were no Ph.D's. at the main lab at the time of the interview. In order to be able to continue to upgrade the qualifications of the personnel, training courses are provided in areas of microbiology, quality control, and serology. Since it is difficult to obtain any professionls with experience in industrial research, significant on-the-job training is required.

The facilities of the lab were described by the R&D managers as being the most advanced for J&J in Latin America, and one of the best in Brazil in any industrial sector, comparing favorably with that of European and American affiliates in the country, and was probably the first to move beyond quality control into R&D programs. The lab is considered so highly that the government sometimes uses it as a reference lab for the setting of standards and for determining necessity of recall of products of other companies. The capabilities of the lab are attested by the fact that J&J has not had to recall any of its own products. The lab is also being used for training of public health technicians on many occasions for several months at a time.

Given the host-market orientation of the affiliate, the major thrust of R&D projects in Brazil is to determine the fit of a particular item from the U.S. company in the Brazilian market and how to manufacture it with local inputs. It has not only found new and different markets for U.S. products, but it has in one instance developed a new product, which eventually was brought into the J&J line in the United States. There is a substantial flow of communication from the U.S. lab to the Brazilian counterpart, which in turn forwards information to other Latin American affiliates. This communication is facilitated by the visits and international conferences sponsored by J&J among its officials. Telephonic communication is also encouraged, occurring at least twice daily to the United States or Europe among the R&D managers.

One of the contributions of the lab to Brazilian development has been the diffusion of skills to supppliers to help them improve the quality of products. The J&J affiliate has frequently given them manufacturing specifications and helped them learn to produce to quality standards and to production schedules. They have even helped the supplier buy specialized equipment to meet specifications and volumes required by J&J. The lab has also been useful to the J&J affiliate in helping to obtain approval of products by government agencis, which see the lab as having such excellent standards that they have been adopted by the government in some pharmaceuticals, adhesives, and sutures. Further contributions to Brazilian development have been made in the improvement of products that have reduced foreign exchange earnings and reduced costs of products, widening the market served. In some instances, annual production costs have been reduced by millions of cruzeiros.

As to the local science community, there is hardly adequate support for the activities of the Brazilian lab. This does not exist either at the universities or among locally owned companies as suppliers or customers. J&J has therefore provided an example for other companies in the development of its quality-control standards and procedures and in the introduction of new products. And the affiliate has always worked to establish close relationships with universities and has supplied some teaching skills on occasion as well as supporting exchanges of professors with American and European institutions.

Unilever Central Research and Overseas Labs

Unilever is a decentralized company manufacturing a wide range of products in the fields of foods and beverages, animal feeds, toilet preparations, and chemicals and chemical products. This case focused on the detergents and foods lines, though many of the comments are valid for other product lines as well as far as R&D is concerned. Once again, the snapshot presented here does not represent the present situation, given the reorganization and restructuring of the R&D activities in the interim.

Unilever has six central research labs located in four countries of Europe, each with a wide range of product responsibilities. (Since this writing, one of the labs has been folded into another, reducing the number to five.) These labs report to the Central Research Division. There are also two "affiliated" laboratories located in the United States and one in India, as well as 29 developmental labs in 25 countries outside of Europe, plus many more development labs on the European continent. Over 7,000 personnel are employed in all types of R&D labs of Unilever. Of these, approximately 4,000 are in the Central Research Labs in Europe, which ranged in size from 100 to 1,500 personnel.

Overseeing all these activities is a Research Division Executive Committee, which centralizes policy control over several of the multiproduct, multidisciplinary, multinational laboratories that serve the entire corporation. The line authority of the Research Division extends to Europe only, excluding the United States and Canada and the overseas affiliates; an Overseas Committe has responsibility for the labs outside the European continent, save those in the United States, which are considered independent. (Although the U.S. labs were visited and are reported in the complete case, they are not reviewed in this summary since the emphasis is on R&D emerging in the developing countries and the labs in India and Brazil provided extensive information.)

The Central Research Labs conduct basic and applied research looking for products to be picked up by the various manufacturing units. These companies are independent in the sense of being able to determine their own objectives and product lines, so the Central Research Labs provide service plus exploration and initiation of new ideas. The Colworth Lab in foods has among the best biochemists, microbiologists, toxicologists, and animal biologists in Europe, with a professional personnel of international renown and a capacity in the food field greater than all the rest of the U.K. R&D capabilities in this sector. It can address any problem in foods of any type; and when combined with the capabilities of the Vlaardingen Labs in the Netherlands, the two laboratories are recorded as a single *country* in EEC statistical tables on food research.

Being both service and initiatory, each lab has some inputs from outside through the marketing division and others from inside through an analysis of potential products by its professionals. A research planning group determines the lab projects that it considers suitable and then divides them among the various central research labs according to the interests of the lab and its capabilities. Over 25 assorted internal reports arise from the various projects, each of which is available to affiliates around the world, as needed. The Central Research activities require considerable coordination, and Unilever officials do not think they could successfully coordinate research dispersed over more than five main centers of activity.

The company development labs are considerably smaller, with personnel numbering just over 100 and down to 10–15 at the smallest. The responsibilities of these labs are to provide (a) developments for new products and the optimum use of new processes for the companies to which they are attached, and (b) technical services and quality control. The budget and manpower are usually split 50–50 between these activities. Budgets for these labs are determined individually by the cost center to which they are attached. There are over 100 development labs plus another 400 companies that have only technical services and quality-control facilities.

A development lab moves to "affiliated" status (coordinated with and partially funded by Central Research) when: (1) It becomes politically desirable to improve the status and image of the lab; (2) the lab is technically sophisticated or is having to support sophisticated products and has developed an adequate capability; (3) the market has become sophisticated enough for differentiated products; (4) government regulations require local testing and evaluation; (5) the lab is highly specialized and of a critical size to support more than developmental research—no less than 6 specialized professionals with say 20 support personnel; (6) its program merits some central funding support. Movement into this status is a question of growth and decision by the affiliated company as well as the Central Research Division. The Research Division is itself not a profit center, but a cost center, and therefore final responsibility devolves upon the operating company that is a profit center and that bears the cost of the affiliated lab, though with some support from the Research Division. One of the problems in moving to affiliated status is that of distance, which makes communication difficult. However, the distance is more importantly cultural than geographic. Cultural differences require frequent visits among personnel to increase familiarity. But coordination is not control or centralization, and Unilever finds the differences in consumer taste and product lines make it inefficient to centralize consumer research.

The status of an affiliated lab includes considerable assistance from the Central Research Labs, including (1) communication of research results and successes and failures of projects; (2) management and trade development in training of technicians; (3) seminars on current issues and research results; (4) supplementary funding for Central Research projects undertaken also by the affiliated labs; and (5) expert assistance in sorting out good ideas from bad and group participation in project designs.

Assistance to the development labs goes through an R&D Application Unit (RDAU), which has been set up at the Vlaardingen Lab to provide support and trouble-shooting in six of the smaller European countries. The RDAU is an effort to provide these manufacturing facilities with prompt service from a centralized source. It does not extend into overseas development labs, but similar services are available on an as-needed basis under more ad hoc arrangements. Unilever considers it difficult to develop an RDAU for overseas, unless each regional area had a separate facility. Therefore, there is a gap in support since this regional set-up has not been established, and particular product capabilities do not exist in each country.

Still there is a ready flow of information that is of great significance and value, especially since it is provided to affiliates at much less than cost; they bear only out-of-pocket expenses of some transmissions. As one of many examples, a technical director in one of the South American subsidiaries wanted to develop dry soups; he requested information from Colworth on how to go about it and received a five-page letter and four extensive technical reports, which put him into the business effectively.

Affiliate Labs: India. The research lab in Bombay, India, is part of Hindustan-Lever Ltd. (HLL) but reports also to the Research Division, making it an affiliated lab. Its purpose has been to work closely with HLL in evolving processes to upgrade indigenous raw materials as part of an import-substitution program and to improve the product line. The R&D center is making it possible for HLL to use several unconventional oils—e.g., castor, ricebran, and sal—in soap making. By 1960, the responsibilities of the lab had expanded along with the product line to include not only soaps and vegetable ghee (a form of local butter fat) but also detergents and convenience foods, dairy products, and animal feeds.

The Chairman of HLL decided to start the lab when he foresaw the need to substitute indigenous products for imported oils. It was tied to the manufacturing plant and became a separate physical unit of HLL only after it expanded to include convenience foods and moved into new facilities in 1967. Despite its separateness, the lab is closely tied into the marketing and product objectives of the company. Success of the lab has arisen from the facts that the peculiar problems of raw materials in India required local R&D; that a local market was large and diverse enough to support R&D; and that the qualified Indian scientists were available. The establishment of the lab has left HLL's competitors far behind, and its growth and development has occurred with substantial assistance from Unilever central labs, helping to learn both background research and consumer-oriented R&D.

Despite HLL's autonomy of operations, its Chairman sees a high value to continued links with the central labs in providing access to international R&D data and processes, training of scientists and exchange of personnel, training in R&D management, and stimulation of professional curiosity through international contacts.

In addition to the R&D lab at HLL headquarters, there is a

development lab at the factory site that remained there when the new lab facilities were built. It now has a function of transferring into the plant developments and adaptations made at the R&D lab. The plant lab is also a training ground for personnel going into manufacturing, making certain that they have adequate technical training.

The total developmental effort of HLL is divided between the factory lab looking at the more immediate problems, and the R&D Center looking at those farther down the road. Responsibility for immediate problems is 80 percent that of development labs and 20 percent that of the R&D Center. Contrarily, any problems that are two years away or longer are only 40 percent the responsibility of development labs and 60 percent that of the R&D Center. Thus castor oil processing was developed in the R&D Center through a pilot plant stage, and the factory helped scale it up. Both units assist the manufacturing plant in interpreting the reports through the Technical Information Service, which indicates problems that are arising in other plants and the solutions that have been worked out. (This description shows the HLL labs to have a host-market orientation, evolving with the growth of the market, operating under virtually complete freedom, and developing products not standardized with the parent.)

The R&D Lab in Bombay is the largest in consumer products in India and the second largest lab of any sort in private industry— that of Ciba-Geigy being first. The lab also has the only toxicology facilities for testing consumer products in India. The replacement costs of its equipment would amount to around £10,000,000 at present inflated prices, and its annual budget has risen from under £300,000 in 1970 to over £600,000 in 1976. Approximately half the budget was expended for projects relating to soaps and detergents, with just under a third for chemicals, around 10 percent for toilet preparations, and the rest for foods and animal nutrition projects. Of the total staff of 215, 42 hold Ph.D. degrees, all from universities in the United States and the United Kingdom. Many research assistants have Indian Ph.D's., with 10 holding master's degrees and 125 bachelor's degrees; 40 are technicians. The dedication to research is illustrated by the fact that one man alone has some 60 publications.

Besides the variety of research projects, the lab is involved in activities related to governmental policies—for example, it helps develop patent specifications, follows up complaints on the use of products, and establishes quality-control facilities in various locations in India. Further, it has participated in the deliberations of the

Central Committee of Food Standards and its various subcommittees concerning the addition of flavor and coloring in margarine and has cooperated with other public agencies dealing with research matters.

Efforts have been made to calculate the cost/benefit of research activities to the company, which had been projected mostly in cost savings, particularly in the use of domestic materials. The lab has calculated that the total research budget has been recovered five times through cost savings, though many of the *potential* savings are not included in the estimate, and no value was calculated for increased sales. The major contribution of course is in keeping the company technically sophisticated and ahead of its competition. It is so far ahead of others that there have been some public suggestions that HLL should give its developments to competitors so that they can remain in the market.

As a contribution to the economic development of India, the substitution of domestic fats and oils has saved considerable foreign exchange and has encouraged scientific and technological employment in the country, including employment in supplier industries. Not only have suppliers of oils benefited but a range of chemicals have also been stimulated including turpentine, citronela, and Indian lemon grass oil for perfume. A large number of chemicals are being produced from local materials, also providing for import substitution.

Another project in which the HLL lab was engaged was the development of prawn farming, which turned out to be successful, but the company decided not to pursue it itself and turned over the results to the government.

Development Lab: Brazil. The Brazilian company was established in 1929, producing at first soap and perfumery. It added soap powders in 1952 and detergents in 1959, acquiring a locally owned company in 1961 that was in powders, soaps, and toothpaste. In 1973 it acquired another Brazilian company in ice creams and still another in industrial detergents. It now produces over 300 products.

The purpose of the development lab related to each of the major product divisions is to substitute local materials for imports. As a result of this activity, imports have been cut from 10 percent of the sales volume to less than 5 percent. The host-market orientation of the lab is seen in the close tie to manufacturing and marketing; in fact, the lab manager reports to the technical manager in each division, who is responsible for all production and product development. About half of the personnel in the lab are concerned with quality control while the rest are concerned with product translation and adaptation.

The personnel at the lab are usually university graduates, but they have been trained in largely theoretical courses, so that they have to be trained on the job for an orientation toward industrial developmental work. This training is done by bringing engineers from Europe to Brazil or sending local personnel to take specialized courses. There are very few Ph.D's. available, so the company is concentrating on Bachelors or Masters, upgrading some of the Bachelors as feasible.

Since much of the work of the lab is in product translation, there is a continuing flow of information back and forth with the central labs. For example, the lab had to scale down the super-fatty ingredients in toilet soap because of the high cost of these ingredients in Brazil. They were successful in reforming the soap to cut the cost without a reduction in performance, and the product is now being adapted for use in Europe also to cut costs. In some cases, the ingredients available in Brazil are so different that European labs are not useful in solving problems. The Brazilian lab has to develop its own capabilities in making substitutions. When the inputs are so significantly different that the product itself might be altered, the lab has the difficult job of convincing the central lab and top management in Europe of the value of using the ingredients for the purpose of cost reduction.

In many cases where the project is concerned with ingredients, the lab works with suppliers who are frequently affiliates of other foreign companies; the parent companies are then able to assist in redesigning the specifications for the use of the Unilever affiliate. Such projects have cut as much as 60 percent in the cost of materials inputs for some products.

The lab is not yet in new product development, but the director of the lab is attempting to get professionals to think of new product form or formulations. Several of the professional scientists have the ability to "blue sky" a bit and to come up with new products for formulation. However, the director recognizes that these efforts might not be rewarded by an adoption of the product by the marketing divisions. One of the problems is that there is inadequate consumer research in Brazil, and it is hard to get the consumer to use the product in ways that can be tested or assessed. The central labs in Britain have helped with some of the design of the consumer research projects, just as they have with some of the product adaptations. (As with the Indian lab, the Brazilian unit is virtually free from centralized control, oriented to nonstandard products for the host-market, and is evolving slowly according to market growth.)

Also tracking with the Indian lab, the benefits of the activities of the Brazilian lab are shown in the reduction of costs of production. The reduction in import costs are not just cost savings but also permit the company to continue to produce and to expand, given the constraints on importation by the Brazilian government.

Given the absence of a strong science community, the company is attempting to help develop that community through exchange of scientists and provision of scholarships as well as exchange of professors, helping them to see the kinds of courses that would be needed for industry-related research.

FUTURE R&D IN
DEVELOPING COUNTRIES

Among the companies studied, Unilever has had a most extensive and long-term exposure in developing countries. Therefore its officials were questioned at length concerning future prospects for industrial R&D in the private sector in LDCs. In assessing future prospects, Unilever officials were asked to take into consideration the states through which the development might occur and the obstacles that exist. Given the heavy orientation to consumer needs, it would appear that substantial research would be required in a number of countries where the market might be expanding rapidly. As might be expected, there were divergent views as to the possibilities and desirability of Unilever undertaking significant R&D operation in the overseas countries during the 1980s. These differences showed up in a discussion of the objectives for such research, the relationship to the Research Division, the scope of research in the developing countries, and the geographic location of labs.

Objectives

Two major objectives appear to be important in the minds of Unilever officials: One is the need to do research on indigenous inputs of materials in a variety of product lines (for example, India and Brazil) and the other is a response to political pressures that the company provide a scientific presence in the host country. In connection with the first, Unilever recognized the need to do more research on foods in LDCs, though it is in the inception stages of its own move in this direction. There was also a recognition that the markets in detergents are rising in LDCs sufficiently to require some local research in that product line. Therefore the greater the demand for products that reflect indigenous needs and tastes, the greater the need for consumer-research and product-development activities. Thus the head of the Vlaardingen lab stated in a speech in 1975, "As consumers start to demand an increasingly varied range of products (in the LDCs), the requirements of greater technological sophistication will also have to be met. It will probably be easier to achieve this if the finishing touch in product development can be applied more on-the-spot. Working at a distance is in such a case no longer efficient."[3]

The objectives of expansion of Unilever R&D in developing countries are to gain new and improved products in LDCs plus better government relations (including with regulatory bodies) and improved relationships with opinion-forming groups in the host countries. The company expects to establish R&D units where they can be commercially effective and also help host countries meet their long-term goals. The company anticipates that companies that do not respond appropriately in R&D in affiliates will lose permission to import technology from the center.

The Unilever Board had decided in 1978 that a greater priority should be given to research by the overseas companies in view of the fact—if nothing else—that less research is done there per pound of profit than in Europe. This would mean more R&D initiated by affiliates, and later more "affiliated labs"—i.e., labs abroad, tied programmatically to the central corporate labs. "Affiliation" means that the lab would be responsible to local management for its objectives and evaluations of results, but to the Research Division for the quality of that staff, the content of its scientific programs, an interchange of information, training, publications, and budgetary control. Still, the cost of the lab would be paid by the local company.

Besides responding appropriately, Unilever officials also recognized that they need to communicate better so that governments understand what they are *already* doing in research. They saw a possible further move to higher level technologies in the production processes so as to provide more advanced experience and training for local workers. Some techniques of production are being developed outside of Europe which might be applicable to the developing countries such as the oil-flotation process employed in South Africa.

Despite these initiatives, Unilever expected to examine carefully where any new labs would be located since it saw no advantage in establishing an unproductive or inefficient lab. It was contemplated that the new efforts would begin in the form of local development labs, tied to local manufacturing companies, and emphasizing quality control and technical service.

Relation to Research Division

Despite the fact that local research activities in developing countries are likely to be tied very closely to the producing companies in each company, the (Central) Research Division will still have a key role to play in their initiation or expansion.

The Research Division has always responded to the needs of the Overseas Committee, but the supporting research has generally been carried out in the European labs. Each operating company in the overseas group can pick up the results that are applicable to its markets. The Research Division was then looking at a couple of countries in each region to examine their needs, but the decentralized structure of Unilever does not lend itself to a *regional* approach to production and distribution, though it might enlarge the market and support further research. Overseas companies are tied individually to the European labs, where all overseas research is synthesized. (Of course, research on some foods would not lend itself to a regional approach, for example, that on babasu in Brazil. Nor can a regional toothpaste be developed for Southeast Asia because of differences in packaging and flavoring among the various countries.)

Some of the officials see a potential conflict between the objectives of the manufacturing companies that will want locally oriented research for their commercial and political survival, whereas the Research Division will want a centralized and coordinated program to reduce costs and achieve efficiency. For example, research on seeds will lead to research on oils that can be used elsewhere than where they are produced and eventually the oils used in products that may be exchanged across national boundaries. Thus the design of the research has to look at consumer needs in several countries, which can frequently be done better at a centralized location.

To some Unilever officials, it appeared quite difficult to set up a *specialized* research unit abroad, unless it was oriented to developing indigenous inputs for a local product. For example, the project on reformed meat would lose some quite significant inputs from the Colworth lab if it were cut off and located in Brazil. There it would not have Colworth's muscle analysis, which is necessary to determine the best ways of cutting and reforming the meat. It would not have capabilities such as electromicroscopy, and it would be far removed from the major markets in which the meat would probably be sold. However, a project on fish farming could be put in India without a great cost to the Colworth program. It would be closer to the market in LDCs and not be as reliant on sophisticated equipment or technology. Even so, India would probably not have all the necessary capabilities.

Unilever is not used to running dissociated labs except in Europe—that is, any lab not tied to an operating company. However, the closer to local operations and the farther the lab is geographically from the Research Division, the less R&D coordination and potential efficiency that can be achieved.

But since corporate headquarters can never be divorced from any affiliate's acts or efforts since the company's total reputation is at stake, there will be a pull to maintain central control or at least surveillance of research in the overseas countries or to tie small, local development labs to an overseas R&D Application Unit—as in Europe—to assist manufacturing in applying technology.

It was estimated by some officials that future research programs in LDCs in detergents would be organized so as to centralize research on common problems or to specialize in LDCs on parts of research—for example, washing at different temperatures. But one of the obstacles to a spin-off of projects into a host country is the difficulty of achieving an appreciation at the center of the differences among the developing-country markets as to product needs and consumer preferences. Unilever has been putting more effort here in the last few years, but it still gets many surprises in the way in which consumers handle foods and detergents. A second obstacle is the necessity to have someone at the other end who has a similar level of professional expertise and is putting the same intensity of effort into getting the results of research applied—that is, someone at the other end needs to be pulling, as well as someone at the center pushing.

The Central Research labs have contributed directly to technical expertise in developing countries through a variety of programs. Programs undertaken through the Food and Drinks Coordination provided some agricultural advisory work, some technical aid, and cooperation with some international organizations working with the developing countries. Within the agricultural advisory work, there were six programs in different countries on growing sunflowers and for testing several varieties in the various countries.

The sunflower research was on the selection, growing, effects of different climates, drooping heads, and the effects of different soils. This research has been given to governmental institutes in Mexico,

Turkey, and other countries. It literally saved a Turkish crop, which had been threatened by disease. The Unilever interest in this crop was to expand the sources of sunflower oil on the open market, and it was still doing research on a dozen varieties. This research was part of a larger program in horticulture on vegetables of different varieties, which in turn required work on soil cultures. It required a team of agrihorticultural experts, chemists, and analytical chemists with wide-ranging capacities.

The Vlaardingen lab has the best experts in this field, and they move around the world for tests on sunflower growing. The results were made public for all interested so that crops will be larger and better. R&D officials have been able to commit resources to this project, despite the fact that it does not directly yield a product for Unilever because it does fit within a wider product-oriented program.

This research has been expanded into rape seed, seeking disease-resistant strains and some with a reduced percentage of undesirable acids. After much research, they have been able to produce a strain without the undesired acid composition by actually cutting the seed so as to eliminate that part which contained the acid. The research covered the responses of the seeds to different soils, climate, the phases of planting, characteristics of the seed, processing problems, and extraction. This kind of work could not yet be done in the smaller labs available in the developing countries. They would have to rely on the capabilities of the central lab, for the rape-seed program required 10 teams and was itself built on a much wider nutritional program.

Advisory programs existed in the early 1970s also for palm oil development in Ghana, on vegetable growing and canning and fruit preserves in Turkey, on mass-feeding programs in India, on vegetable growing for frozen-food exports in Morocco, and on agricultural support for industrial development in the Mekong Delta. Technological assistance stemming from research in the central labs included the use of vegetable protein in institutional food in Iran, extruded vegetable protein as a rice substitute in Indonesia, soya milk processing in Sri Lanka, oil refining in Liberia, integrated meat development in Kenya and Botswana, fishery development in Venezuela and Peru, and protein fruit development in Nigeria. These last four were in cooperation with the Food and Agricultural Organization (FAO).

Agricultural programs in the LDCs are dominated by the FAO, AID, and the U.N. Development Program (UNDP), consequently Unilever has restricted itself to rather small-scale programs. However, it has been working on an air-inflated greenhouse so as to help diversify agricultural output. This type of program has been welcomed by LDC governments. Where the company has had an interest in large-scale programs, it has had to move slowly and through governments since commercially oriented agriculture would alter the culture of a country.

The establishment of an R&D Application Unit (RDAU) for Latin America and other regions will be influenced by the size of the market, similarity of markets, and the ability to be supported by a central R&D lab. Such a RDAU could not be put in a development lab within a Brazilian company and at the same time service all the other countries from Brazil. There would be too much jealousy and suspicion for this to be successful. It has to be separate from the operating companies so as not to seem partial to any one company in the system. (Unilever did have all its development work for Scandinavia in Sweden, and the other countries distrusted it.) Without the backing of a Central lab, an RDAU for Latin America would not have sophisticated service that would be needed to make it effective.

One lab official argued that an RDAU for developing countries could be established at Colworth, oriented basically to problems of processing foods; but even this would leave some difficulties uncovered because of the distance from the operating companies. More physical distance will make it more difficult to handle than in the European RDAU, but RPG chairmen are, among them, now taking as many as 40 trips a year to overseas companies, and younger professionals are coming up who are willing to assume similar responsibilities. The company would send some of its best personnel, as it did in Brazil, to help establish these development labs; it would expect to lose them to the operating companies—at least for awhile.

Development of an R&D presence in an LDC is illustrated by Unilever's activities in Nigeria. The company was able to help the government find food substitutes, easing its nutritional problems, but without direct profit to the company. It was also permitted to produce toilet preparations, which were profitable. By 1978 the affiliate needed a small lab of four to ten people to keep production going; once established, its activities would provide employment for scientific and technical personnel. These personnel will be involved in quality control and some research on mineral oil substitutes. Later they will get involved in problems of different types of hair and hair styles. The types of hair and styles are not the same even throughout Nigeria, reflecting different customs, but the company will still have to produce a single (undiversified) line of products.

LDC Research Programs

The first program stage for an LDC lab is the establishment of quality control related to manufacturing. Gradually, depending on the market and the availability of manpower, product development can be begun. There is, however, no necessary functional relation between quality-control responsibilities and functions of a development lab. Very small operations in research can be begun on food simply by one professional in a kitchen, mixing powders or dehydrated soups or sauces or flavorings. However, if this small operation is tied into a larger system, quick answers can be provided to new questions, preventing redesign of the wheel. Such exchange of information requires that the person in the kitchen be a sophisticated professional so that he can ask the right questions and understand the answers.[4] He also needs to have knowledge of how to adapt the answers to the local market and therefore needs to pay attention to local market needs and government objectives.

Given their separation from central labs, labs in the developing countries need to be able to make quick responses on their own; the problems they face change as governments shift their economic and social welfare objectives. Governmental requirements as to new types of products, local content, and safety also alter definition of the research problems. This makes the establishment of effective programs in developing countries labs extremely difficult because most R&D labs are "programmed" for yesterday's problems, and top management for day-before-yesterday's problems, in comparison to what is needed now or in the future. Present R&D managers have a difficult time envisioning the future, changing needs in developing countries.

Specialized labs in LDCs would be very difficult to orient and control. Specializing in cloves in Zanzibar would be difficult to staff and operate because it would need an understanding of markets (which are outside the country) and a sophisticated expertise in problems of growing agricultural products. It would also need an efficient scientific manager, plus dedicated personnel. This is not to say that it cannot be done, just that it is most difficult.

The product development effort in Indonesia is oriented toward use of local materials and to meet nutrition problems. Java is growing soybeans for personal consumption only, and there is no commercial farming. The shift to commercial will probably be made by company efforts on Sumatra so as not to disrupt Javan customs. To accelerate the necessary shift, research programs would be needed on soils, processing, distribution, and use of soybeans, probably in cooperation with the UNDP and FAO. Many Javan university research teams are working on soybeans but are not significantly oriented toward commercial uses. Unilever could assist in reorienting the research. Any effort it would put in should be tied to the operating plant in Indonesia for its support and for greater realism in the research.

To get Indonesian support, it might also send experts to talk with officials on quite different projects in which the government itself is interested such as arterio-sclerosis, which is rising in wealthy groups in Djakarta and is found among ministerial officials. Unilever also has a research program on growing palms to produce oil, aimed at improving plantation efficiency. Its research is based on an analysis of vegetable tissue. It is giving the results to the government, and it still has a long-term program in the subject.

The recently established lab in Turkey suggests the scope of a new facility in an LDC. The objectives of this lab are fourfold:

To bring out and submit to public benefit the unused resources of the country.
To establish links between research institutions in universities, industry, and government.
To establish a laboratory for research on the growth of oil seeds.
To increase local contributions to research.

The projects initially undertaken included assessment of cooking procedures and industrial resources, utilization of forestry products, assessment of perfume plants, and synthesizing perfume components. Several studies were undertaken with reference to oil seeds and nuts—sunflower, rape, ground nut, tobacco, tea, grapes, and pistachio. The trials with the new sunflower seeds are in cooperation with university faculties, agricultural research institutes, and the sugar company of Turkey. The demand for ground nuts (peanuts) has increased for snack foods, and research is being expanded to increase the acreage planted in ground nuts. Tobacco seeds are normally wasted by the peasants in Turkey, but they yield an oil useful in cooking and in industry, and the cakes can be used for animal feed. Tea seed is also a source of cooking oil, yet it has not been used in this fashion in Turkey previously. The cake from tea seed also contains a component called saponin. Grape seeds yield about 13 percent oil and are now simply thrown away in the manufacture of wine and other products. Pistachios grow wild in the country, and their use for oil is very small presently; the research is aimed at expanding the potentials of this nut.

The Turkish program also includes research on the many scented plants grown in the country, particularly around the Aegean and the Mediterranean regions. It will also examine the availability of raw materials from which synthetic perfumes can be made.

These are the initial components of the research program. All are directed at increasing available supplies of products needed in the country, based on indigenous materials.

Location Criteria

Even after it has been determined that research shall be undertaken in developing countries, it cannot be located in every one of the countries, nor to the same extent in each of the labs of affiliates. The criteria for location will be related to the existence of manufacturing operations, the availability of support facilities and easy communication, and governmental restrictions.

Operating Company. The primary locational criteria would be the existence of a Unilever affiliate already manufacturing in a given country. There is a need for local administration and a local source of funds to take care of the services required by the personnel in R&D. An isolated lab requires too much administration and support services to be economical. Also there needs to be a visible application of what is being done and an urgency to it; otherwise the lab becomes academic in its orientation. The structure and product line of the company would dictate the type of research undertaken by the lab, and the eventual size of the lab will be determined by the size of the markets served by the operating company. This market does not have to be merely national; it can include exports.

Facilities Available. Another criterion is that related to the facilities available to the lab. Not only must there be administrative support but also adequate communication links are required back to Unilever central research labs. If these ties do not exist, the personnel at the center who are responsible for success of the overseas effort cannot follow through. In addition, there is a need for a pool of scientific personnel—both Ph.D's. and others with R&D experience, even if this has been gained abroad.

Further, scientific work requires sufficient physical facilities so as not to add to the frustration of the scientist. This means good sanitation, good housekeeping, adequate supply of inputs, and prompt mail and telephonic communication, not only locally but internationally. The need for such facilities in a development lab are in a sense greater, or at least different, from those of a Central lab. For example, a small Unilever lab needs greater access to outside data banks than do the larger labs, which can collect their own data inputs. For many developing countries, access to the central data banks is difficult because it has to be done electronically and by satellite. Facilities for this simply do not exist in many LDCs, but if these various facilities are not appropriately supplied, the cost and frustration become so great as to make it advantageous to shift the operation back to the central labs.

Although training is necessary for the scientists at an affiliate's lab, Unilever's experience is that there is a danger in training them highly. Training of development personnel beyond the first university degree gives them an image of themselves that prevents them from working on the factory floor where it is necessary to translate research results into production. Training is also costly beyond the necessary levels; chemistry at lower levels is a simple mixing of elements and ingredients once it is known what to mix. If the formula is provided from the center, no highly advanced knowledge is required to do the mixing at the lab.

Finally, in the view of several Unilever officials, a higher priority should be given to support from governmental institutes, which can provide substantial basic background research. And governmental support for scientific education will be necessary. For example, Turkey has had a long history of education and science based on a close relationship to the German system and relying on German personnel in the past. Consequently, it has a good nucleus of technicians, on which to build developmental work.

Government Requirements. A proliferation of governmental requirements that R&D activities be set up would pose serious problems for Unilever's operations abroad. The company simply could not put R&D activities in every country where it had a manufacturing operation. However, *some* scientific and technical work will be required at each such location simply for the purposes of technical services and quality control.

If the company is pushed beyond this stage by government regulations, in order to achieve higher levels of scientific research, it could cut off a few pieces of research in a country such as Brazil, hiring less than ten professionals on a quite specialized activity, but, as discussed previously, this raises serious problems also. The selection of a country in which to set up a specialized lab would depend on the leverage that such a lab would have not only on government relations but on the future commercial development of the company and its future role in Unilever research.

One of the more difficult problems in determining location of R&D activities in the future will be the government's attitude toward joint ventures. However, from Unilever's viewpoint, each case will stand on its own. In Portugal there is a small joint venture in which the partner wants all the R&D assistance that he can obtain. Assistance was initially offered free by Unilever, focusing on methodology, problem identification, and research design. Once the partner saw the usefulness of these activities, the venture began to create its own development lab at its own expense.

The willingness to provide extensive scientific and technological support overseas depends on the type of partner and his ability to use the information outside of the partnership. In India, for example, the partners are private individuals, holding shares sold publicly, and so long as Unilever's ownership remains above 50 percent, the Overseas Committee of Unilever saw no significant effect on the R&D program. If it drops below 50 percent, the relationships are likely to be reassessed.

The alternative of separating the R&D lab from the manufacturing company in order to retain 100-percent ownership of the lab alone is not attractive to Unilever since it would divorce lab activities from the production needs of the manufacturing company. A related problem would arise in the ownership of patent rights and the charges to be made for the assistance given the lab in the locally owned joint venture. The scope and nature of cooperation would be changed fundamentally, but Unilever has not yet prepared a policy for this contingency.

Summary

Summarizing the many conversations with Unilever officials, the following maxims for creation of effective development labs in LDCs can be set forth:

1. The establishment of an R&D lab in a developing country requires a large enough business locally to support a lab of some 30 scientists, ranging over several products.
2. Research problems should be sufficiently peculiar to the country so that they can be more effectively tackled through the local lab than at the center.
3. A national science community such as university institutes, governmental institutes or numerous R&D labs in other companies should exist to support the local unit.
4. An adequate number of qualified scientists should be available.

5. R&D labs should not be established unless there is an *economic* basis for doing so and a *commercial* benefit to the company— never simply for public relations. Without a related commercial market, it is difficult to determine how large to make a lab. How should one be set up that is specialized and unrelated to the commercial operations in a country that is simply insisting on R&D? On the other hand, there are obvious advantages of setting up a local lab to develop indigenous products for a growing market and to help in their processing. Good public and government relations would come out of sound work by the lab; but it should not be set up solely for this purpose, for it could never demonstrate to government officials its usefulness.

6. Management of the operating business should have an understanding of R&D and a belief in its long-term results. R&D has to be conducted even in lean years to provide proper phasing between R&D output and commercialization of improved or new products.

7. Other parts of the operating business should be sufficiently sophisticated so that the results of R&D can be utilized effectively. Thus engineering, production, and marketing departments have to be educated to accept and use R&D efforts. This provides the pull necessary to introduce R&D into the system.

Given these conditions, Unilever officials consider it logical to follow the same road in LDCs that it has elsewhere, fitting R&D activities to commercial needs and expanding with the market.

NOTES

1. The six Japanese firms are not included in the analyses presented in this chapter because of their total lack of foreign R&D activities.
2. The management style scheme employed in this study was based on Henri DeBodinant, *Influence in the Multinational Corporation: The Case of Manufacturing.* Unpublished doctoral thesis, Graduate School of Business Administration, Harvard University, 1975.
3. Professor J. Boldingh, October 13, 1975.
4. All officials indicated that if the communication had to be over 1500 miles, it would be hard to handle, especially because the "person in the kitchen" would be very busy trying to carry out his activities and to decipher and apply the information given him. The further away the lab is geographically from the center, the harder it is to keep communication effective simply because visits would be cut to the bare minimum if not to zero, and it would be difficult to narrow the gap in perception.

Chapter 2

Pharmaceutical Programs in Tropical Diseases

In discussions with companies as to the types of case illustrations that would be useful in demonstrating the problems of R&D generation and application, one of the pharmaceutical companies suggested that the role of the industry in mitigating tropical diseases would be an instructive study. These officials suggested that an examination of the process of R&D and its translation into useful drugs and delivery into the field would help show the limited, but critical, role of R&D while illustrating the multiple decision groups involved in the process of invention, innovation, and application of drugs. Since the other case studies were limited to the R&D function, it appeared useful to carry one case study further, through the process of testing and commercialization. Several pharmaceutical companies were approached, and the study was shown to be feasible and desirable because of the concern being expressed at the time (1977) over the pressing need to mitigate several tropical diseases. Like the other case studies, the results shown here are dated; they apply to the years 1977–1978, and many changes have occurred that prevent the descriptions from being current. However, like the others, these observations remain illustrative of R&D problems and processes and help, thereby to focus the issues on how to translate inventiveness into commercialization.

The problems arising from the pervasiveness of disease in developing countries have been the focus of concern over the past several

The larger study of which this is a summary was written by J. N. Behrman and published by the American Enterprise Institute, Washington, D.C., 1980.

years. The roles of the R&D programs of pharmaceutical companies have come under scrutiny with the intention of stimulating greater efforts by the companies and a higher research priority. Much in the R&D programs is applicable to LDCs because many of the diseases found in developing countries are similar to those in advanced countries and have responded to drugs developed in the latter. And cooperative efforts have led to improvement of general health conditions in LDCs. But as a result of these successes, greater attention has been paid recently to some of the more pervasive and intractable parasitic diseases found in tropical countries. The World Health Organization (WHO) has embarked on a Special Program in Tropical Diseases, and concern has been voiced in the U.S. Congress that more should be done to relieve the large masses of people in poorer countries of some of the debilitating diseases to which they are still subject. The six diseases on which attention has been focused are malaria, schistosomiasis, filariasis, leishmaniasis, trypanosomiasis, and leprosy.

Malaria causes more than a million deaths a year, mostly in tropical Africa. Schistosomiasis exists in over 70 countries, infecting some 200 million people, but usually in conjunction with other diseases. Filariasis attacks over 250 million people, taking a number of forms, including swelling of the limbs and blindness. Trypanosomiasis is a sleeping sickness found largely in Central and South Africa; some 35 million people are exposed to the disease, with 10,000 new cases occurring on all continents of the world except Australia and Antarctica, though it is less evident in the United States, Southeast Asia, and some areas of Africa. Some 11–12 million cases of leprosy exist, principally concentrated in tropical countries.

Many of these diseases are related to poor sanitation and health conditions, including the absence of potable water and good nutrition. Some drugs are available for each disease, but they are used in varying intensity around the world depending on public health facilities, governmental funds support, cultural habits such as the use of self-medication, the availability of doctors, the practices of doctors prescribing medications, and the availability of hospitals. Even with better sanitation habits, however, those with the disease presently will require drug therapy to remove the causes and symptoms of the diseases, and for this purpose the pharmaceutical companies have been urged to do more to develop and provide drugs that are effective, safe, inexpensive, and easily dispensed. The ideal formulation would be a one-shot vaccine that provides permanent protection against each of the diseases, but the industry is far from obtaining this solution for these particular diseases.

Despite the pervasiveness of these diseases, not all the governments in countries where they exist have dedicated significant resources for the purchase and distribution of drugs nor even for information to be distributed as to the availability of medication. Given the poverty of most of the individuals infected with these diseases, there is an insignificant market demand for most applicable drugs (with the notable exception of drugs for malaria and schistosomiasis in those countries where the government is the purchaser). Consequently, an investigation of potentially useful compounds by the pharmaceutical companies is a dedication of resources to social needs rather than to a direct market demand.

Several of the major companies have maintained continuous research and developmental efforts in chemotherapy for tropical diseases. Obviously, the companies have done a great deal in removing health problems in developing countries through their work on diseases prevalent in advanced countries but which exist also in the developing countries such as smallpox, yellow fever, measles, TB, infectious and metabolic diseases, and a number of health problems such as hypertension, heart attacks, and contraception for family planning. The six tropical diseases are, however, inadequately attacked or mitigated at the present, and greater resources will have to be committed to them if they are to be contained.

The WHO is trying to stimulate the companies to increase their attention to these diseases. The WHO program seeks "to develop and apply new diagnostic methods, chemo-therapeutic agents, and vaccines especially suited to prevent and treat the diseases of the tropics and the countries affected by them" and "to strengthen research in the countries affected by the diseases by training of scientists and technicians in the relevant discipline." It is trying to secure cooperation of the companies in its program for the discovery and production of adequate drugs and their distribution. Six research task forces have been set up for the diseases in collaboration with existing university, company, and independent laboratories. In addition, training programs will be developed for scientists in tropical countries, and programs are being directed at field distribution drugs. However, direct involvement of host-country governments will be required, and substantial funds will be needed. Present efforts to use existing drugs would need to be expanded, however, before it was evident that host-country governments were placing a sufficiently high priority on the control of these diseases to warrant expanded cooperation from international financial institutions or from the pharmaceutical industry itself. Included within these governmental programs will have to be an emphasis on the creation of adequate local laboratories to establish cooperation with the international pharmaceutical companies. It is a purpose of WHO to begin the process of creation of such laboratories throughout the Third World.

ACTIVITIES OF INTERNATIONAL COMPANIES

At least 14 of the major pharmaceutical companies are engaged in R&D on tropical diseases. Among these, industry officials consider eight to be committed to continuing research in this area, while others are peripherally or indirectly involved. Since there is neither a strong market demand nor a likely commercial pay-out for drugs related to these diseases, work in this field must be justified by the companies on other bases. For some it is service; for others, full application of existing capabilities; or, hope for a significant breakthrough. In many instances, the R&D programs are continued simply because of the personal interest, dedication, and experience of the professionals in the laboratories.

Beside the development of drugs for the six parasitic diseases, the international companies assist in containing these diseases in a number of ways: support of delivery systems in both private and public sectors for the relevant drugs; cooperative projects in research on potential compounds; cooperative training of scientists in developing countries; and performance of clinical and field trials in the host countries.

Drug Discovery Efforts

In seeking drugs applicable to the six diseases, several have been developed over the past years that are variously effective: There are ten different drugs available against malaria; six useful against schistosomiasis; possibly three against filariasis; probably five for trypanosomiasis; at least five for leprosy; and three for leishmaniasis. Some of these are of long standing (forty years or more) and others are quite recent.

For some diseases, control is best accomplished directly through the patient and others through the carrier or vector. Thus leprosy is controllable by chemotherapy while leishmaniasis and Chagas' (a type of trypanosomiasis) are controllable through attack on the vectors (parasite carriers). Malaria, trypanosomiasis, schistosomiasis, and filariasis are now controllable through both vectors and chemotherapy; a campaign against them requires attacks on both simultaneously.

In order to find new drugs for these diseases, companies have various programs for screening compounds they may discover. This screening is sometimes done on an ad hoc basis rather than under a continuing and full-scale program. Some of the screening is done in the company laboratories; some is farmed out to university or independent laboratories. Other companies find a spin-off from research in veterinary medicine into compounds for human use such as in trypanosomiasis.

A complication in the screening of drugs for parasitic diseases is that one drug may work for a given type of the disease, though it may not work for a different form of the same disease. For example, schistosomiasis has three different forms, which has required three different drug formulations for their control up to the present. In addition, any new drug formulation will be different according to the way in which it is applied to the patient—oral or intramuscular or a vaccine providing immunity.

In addition, several different parasites may produce a single type of disease—for example, five or six different worms cause filariasis. Further, the parasitic diseases require the elimination of a living organism from the human body. This requires killing one organism without damaging the host—not an easy task since the cellular composition of parasite and human host is not greatly different.

Despite the difficulties, several companies have persisted in research in these diseases. In the United States, Pfizer is concentrating on schistosomiasis; Parke-Davis/Warner Lambert worked on all six of the diseases in the 1960s but has concentrated mostly on schistosomiasis and malaria in more recent years; Janssen (Belgium) has been working on schistosomiasis and filariasis with limited programs in leishmaniasis and trypanosomiasis. In Switzerland, Hoffman-LaRoche has been working on some of the six tropical diseases since 1953 and now has projects in all of the six except leprosy; and Ciba-Geigy has developed an effective drug against schistosomiasis and another against leprosy, but has dropped a program on malaria because of lack of success. The Wellcome Research Laboratories (Britain) have been in tropical medicine since 1913, concentrating on five of the parasitic diseases, excluding leprosy; some 15 to 20 percent of its total research budget is in tropical-disease projects; it has the drug of choice for leishmaniasis. In Germany, Bayer has been working on malaria, filariasis, and schistosomiasis; and Hoechst has been working mainly on malaria and filariasis, having dropped its program in leprosy some ten years ago. Lepetit (Italy) discovered and has licensed to others a successful drug against leprosy.

As a consequence of these efforts, Bayer, Ciba-Geigy, Pfizer, and Merck/Darmstad have drugs used for schistosomiasis; but there is no really effective drug against filariasis as yet. Drugs against malaria have shifted back and forth in response to the development of resistance by the parasites. Nothing new has been developed for trypanosomiasis over the past forty years.

Not all of the companies are pursuing chemotherapeutic agents in each of the diseases, and their decision criteria depend on their assessment of the following factors:

The medical need for a particular drug

The market potentials for its sale

The present capacities and capabilities of the laboratory, including those it might acquire easily from outside

The probability of success of the project—that is, its experimental feasibility, including the existence of disease models

The cost of the project in money and manpower and in the necessity to drop another project

The existence of competition in the field

Expectations of government provision of enforceable patent protection

Assessing these factors, a number of companies have decided simply to stay out of the tropical disease area while others consider that their special capabilities and interests require them to maintain some projects in the field. Some companies are responding to concerns expressed by WHO and governments that efforts should be dedicated in this direction despite the current absence of market opportunities. Some companies have gone into research for compounds against a particular disease only to be frustrated through failure at one stage or another of the process; or encouraged by the achievement of clinical success only to find little or no willingness on the part of governments to purchase and supply the drugs where needed.

Some companies have stayed out of the field because of adequate attention paid by others such as the Walter Reed Hospital Program in malaria. Several companies have determined that their research capabilities (including managerial effectiveness) are already stretched beyond the optimum, and they cannot justify substituting work in tropical diseases for existing projects. Even so, if there were a substantial market for applicable drugs, laboratories would probably go out and buy expertise or contract out projects with independent laboratories so as not to miss out in a lucrative area. Since this is not the case, there is no pressure on them to expand in this way. Sometimes, however, a company feels that a particular problem area is important despite the absence of the market. Also professionals will sometimes press to undertake a project simply to assuage their individual curiosity as to whether or not something can be found.

In seeking appropriate drugs against the six diseases, some contradictions arise between the criteria of an "ideal drug" as seen from the standpoint of a host government compared to those applied by a company. The ideal drug for a company is one that would be given orally, in a single dose, that is highly effective, stable, simple, relatively inexpensive, active internally, and patentable. If a company can find such a drug, it will push to produce it regardless of how narrow or specific the market. When these characteristics do not exist, company decisions are based on trade-offs. These trade-offs do not match necessarily the goals of governments. Thus if a drug has to be complex to be effective and requires high technology, it is likely to be more expensive than governments wish. Also companies are looking for drugs that are patentable and therefore protected against competition, but governments are looking for generic drugs that can be produced by a number of companies and at reduced costs.

Despite the difficulties, companies are making substantial efforts to find antiparasitic drugs and will continue to do so. Some companies even have programs in immunology, despite the fact that immunization against parasites sems exceedingly difficult. One of the problems is that of timing—the parasite must be killed before it harms the patient. This means that the invader has to be paralyzed and kept immobile so that the body can generate attacking forces in that spot. Attacks against multiple-cell parasites require that the research go back to fundamental biology and chemistry to find out how the invaders work and what chemistry can be relied on to affect key cells in them. This process of discovery will be exceedingly costly and time-consuming. Estimates by lab officials range from five to ten years for a vaccine against malaria and ten to twenty against any of the other 6 tropical diseases—if they are ever discovered! Even so, Roche, Ciba-Geigy, and Wellcome are among those that have programs to develop vaccines against parasites.

Assistance in Delivery Systems

In developing countries, only about 5 percent of the population is reached through normal marketing channels for health care. Therefore the government is frequently the only or major purchaser, distributing through public health services. Many of the pharmaceutical companies have assisted in improving these and in cooperating with public health authorities in food production and preparation, nutrition, sanitation, and various educational efforts such as audio-visual materials to increase awareness of particular diseases and the way in which the vectors can be controlled.

Various experiments are being conducted in programs of several countries to determine whether treatment can be begun early for diseases—for example, one in Nigeria treats children up to five years old for malaria, using lower dosages; those between five and ten years are being treated in school; and those eleven through sixteen years will be treated only when they have a diagnosed infection. Each province in Nigeria is carrying out its own part of the program with considerable assistance from a pharmaceutical company.

Some experience is being gathered in several countries in "total program" implementation and necessary administrative services so as to assist the government in development of sanitation facilities, vector control, and education of the people subject to infection. However, to date, few governments have shown either the interest or the ability to adopt such a total orientation or to manage it effectively.

Cooperative Training

Several international companies are already engaged in assisting scientists from developing countries (whether or not attached to a laboratory) in obtaining skills as professionals or technicians through bringing the individuals into their company laboratories or working with them in their home areas. Both of these approaches have problems, and some of the companies have simply refused to accept trainees for reasons of secrecy or frustration with the results.

The three major pharmaceutical companies in Switzerland set up the Basle Foundation on Tropical Diseases, which has helped train young researchers and public health officials from LDCs. It has trained Africans for three to four months in the bush, with the result that it has improved radically the sanitation in some rural regions. The companies consider training in research techniques in

Switzerland to be somewhat counterproductive because the scientist gets the idea that sophisticated equipment is necessary for research, but he will not be able to obtain such equipment when he returns to his own country, and it would not be really applicable to local problems. Therefore the foundation takes trainers to the local environment to create a proper orientation.

One company began its training through the institution of quality control on imported drugs in the host country. It staffed a school for technicians in Indonesia with the cooperation of the government, which provided some teachers and equipment. Personnel were trained for two years, and then the operation was turned over to the government, which continued to run it under the direction of one of the former students. Successful programs in this direction are few, however, because of a relatively low priority given to these efforts in the developing countries themselves.

Cooperative Research

To help build up research capabilities in the developing countries themselves and thereby an appreciation of the local needs and possible contributions from the international companies, several companies have developed cooperative research programs in developing countries. One such program was in the former East African Union of Tanzania, Uganda, and Kenya. Another was in the Philippines; a third in Latin America; and a fourth in Bangladesh. Cooperative screening programs have been conducted with independent and government-supported laboratories in Brazil by several countries simply because the scientists there are advanced and are themselves working on parasitic diseases. There are only a few countries in the developing world, however, that are ready for such laboratories because of a basic lack of curiosity and scientific orientation on the part of educated groups in the country.

The most extensive cooperation of this sort is screening of compounds. A company sends compounds to an independent laboratory or a university laboratory for screening against various diseases—against screens either in vitro (glass tubes with appropriate chemical solutions) or in animals. Additional cooperation occurs through clinical testing, but one of the critical elements here is whether or not the foreign partner will follow the protocol (procedures for carrying out tests and recording the results) established by the international company.

Despite the difficulties, many of the international companies will continue to seek ways of cooperating with laboratories in developing countries since they recognize that scientific capabilities must exist there to be able to appreciate mutual needs and potential contribution of the companies.

Cooperative Trials

In order to get the drugs used effectively, they must pass clinical trials for safety and efficacy and then go into field trials to determine whether or not they can be applied in the setting where the disease is found. The clinical trials frequently have to be made in the host country, following regulations laid down by the government in order to achieve approval of sale. The field trials are required to make certain that the patients will come for the number of doses needed and that they can in fact respond appropriately to the dosage level and form in which the drug is provided.

Since the clinical trials require following very precise procedures (protocols), close cooperation is needed between the research laboratory of the international company and the facilities in the host-country hospitals or universities carrying out the trials.

Field trials are even more difficult and extensive and are more important for the public health unit (or the Ministry of Health) because they are not concerned with statistical validation in the small samples but with adequate efficacy in a real situation. The officials cannot justify large sums of public health funds unless success is achieved in the field. Assistance of pharmaceutical companies in such trials usually takes the form of the provision of the drug either free or at very low cost, plus some supervisory personnel to assist in the trials and in the recording of the results.

The research and developmental effort from drug discovery through the delivery system approaches a "total program" orientation to health needs, which is what the international companies argue is necessary to meet the problems in parasitic diseases. This total program requires not just the discovery and production of a drug but the adequate testing of its use in particular environments, the development of delivery systems, educational programs, vector control, and improved sanitation conditions to remove the incidence of infection. Such a total program orientation will mean increased attention to the problems of developing countries from several departments of an international company (R&D, production, and marketing) and a higher priority in the developing countries, plus support from several departments in each government. For example, in one country, agreement was obtained from the military department to conduct field trials on soldiers who came from various areas of the country; this provided a substantial sampling of the universe of people infected. Health and social welfare departments are also usually involved in such programs as may be the department of labor.

To help inculcate and support a total program approach, the companies consider that the support and persuasion of the WHO will also be needed as well as that of international financial institutions who can provide necessary funds.

COOPERATION IN LDC PROGRAMS

As an indication of the support that international companies can give to public health programs, the research turned up a number of instances in which the companies have either assumed direct responsibility for work in the developing countries or have cooperated closely with the governments in forming the support systems and carrying out campaigns against particular tropical diseases. For example, Ciba-Geigy carried out, over five years in the late 1960s, a campaign against schistosomiasis and the snail vector in an area in Madagascar. The program was conducted under an agreement with the government and was later turned over to the government for its continuation. The program was clearly successful in eradicating much of the infestation of snails in the area and in reducing significantly the incidence of the disease in the populace. However, two years after the government took over the program, the area was reinfected and the snail population was back to prior levels. Many of the problems in the project had little to do with the drugs but rather stemmed from social and cultural conditions. The people feared the incursion of a foreign company and were not used to taking advice from doctors who generally gave very little attention to their patients; there was substantial migration in and out of the area, requiring careful censustaking and record keeping. The general health of the people and their food habits were not good so that the causes of any given results were not always identifiable. The project included over 23,000 people of which over 80 percent were checked for the disease and provided medicine, as needed. The conclusion drawn from the project by those responsible was that the most useful contribution to disease eradication was in improving the sanitation habits of the people rather than in drug therapy.

Ciba-Geigy also has conducted training programs in Tanzania to help provide rural paramedical personnel who became aides in the dispensaries and hospitals that treat tropical diseases. The training program is directed at sanitation requirements as well as efforts to mitigate the endemic tropical diseases. The results have shown that habits can be changed in terms of use of water for drinking and cooking, use of toilets, storage of foodstuffs, smoke outlets and ventilation, protection against rodents, control of sand fleas, and the use of the mosquito netting. The trainees have to be familiar with 44 serious diseases plus some 20 minor ones. On graduation, the students generally become the chief of a health center in a rural community. Some 300 such centers are to be created by the year 1980. Ciba-Geigy has also assisted in a cooperative program in Indonesia, begun in 1975, to stop the spread of TB.

Janssen cooperated with the Philippine Health Department and the U.S. Naval Research and Development Command in field trials to determine the effective use of a drug against nematodes. Nearly 100 percent of the population in Northern Luzon carry intestinal parasites, and an effort was made to conduct mass eradication. Not only are the diseases caused by these worms as significant in some areas as the six tropical diseases, but they are so debilitating (consuming food ingested by the patient) that they make individuals easy prey to still other attackers. Special programs were mounted for the determination of the incidence of the disease and the dosages required, plus dissemination of information to get the cooperation of the population. Efforts were also made to change the cultural habits of the people and their beliefs as to improve sanitation and health, thereby reducing infection.

A drug developed by Parke-Davis was used in campaigns against leprosy in New Guinea and Micronesia in the late 1960s. A high prevalence of leprosy had existed in the areas for nearly twenty years, and the entire population of three villages was given regular injections of the drug. In one area the program resulted in only six new cases arising over a three-year period, compared to an expected 33 in the absence of treatment. The disappearance of skin lesions was accelerated after treatment began, and high cure rates were obtained when treatment was given upon the inception of the disease. The success of this program was enhanced by the fact that this particular drug required application only every 75 days rather than the more frequent oral dosages used in the past. The doctors concluded that no other treatment would have been feasible in this particular area and that the success was intimately tied to the formulation and dosage procedures permissible with this particular drug.

Pfizer developed a drug that was employed in the Brazilian campaign against schistosomiasis. It was found to be effective against the form of that disease found in Northern Brazil. The Brazilian government had begun efforts to control the disease at least fifty years ago, and new techniques and methods of control had been actively discussed over the past thirty or forty years, but the drugs in existence were not adequate.

Pfizer's new entry was developed in 1970, and five years of clinical and field testing were conducted before the drug was put into public use. Clinical trials had been undertaken in Brazil quite successfully; but the Ministry of Health could see very large expenditures as necessary, and it conducted local field trials to produce thorough documentation that the drug would be effective among peoples located in widely dispersed areas. The Superintendency of Campaigns (SUCAM) of the Ministry of Health conducted its own field trials with a 96-percent cure rate. Officials of SUCAM stated that, as public health officers, they have a "concept of acceptable risk" since they are thinking in terms of the community as a whole, while the academic researcher looking at clinical tests emphasizes the protection of each and every individual.

To get a concerted effort behind the total program, SUCAM sponsored a national conference in 1977 bringing experts together from around the country to listen to the results of tests. With broad gauge support, an allocation of Cr. $1.75 billion was made by the government to bring the disease under control. "Under control" means the reduction of reinfection through the reduction of egg production in humans, which in turn reduces the effects of the disease and the incidence of serious complications from the disease. To break the cycle and maintain control, a program of sanitation and improved water use was necessary, keeping people out of the water in which the snail enters through the skin. An educational program was necessary as well. If any of these get "out of phase," control would not be achieved.

Therefore, a multifaceted program was mounted including (a) the reduction of people infected through the use of chemotherapy; (b) motivation of communities to adopt improved sanitation; (c) reduction of snails through use of molluscicides since the snails are host to the parasite in waters which are breeding areas; (d) avoidance of new sources of disease in water and irrigation projects; (e) avoidance of migration of the disease; and (f) an integration of the program with other government agencies in the construction of adequate water supplies, sanitary sewers, laundries, public bath rooms, and the improvement of home sanitation conditions.

This program was in effect a total effort in which an additional Cr. $800 million was required for sanitation projects in the endemic areas. At then-existing rates of exchange, this amounted to U.S. $180 million over a three- to four-year period. The program would cover eventually over 2,000 municipalities in rural areas throughout the entire country reaching between 8 and 10 million infected persons.

The program required the concentration of activities within specific villages and rural areas beginning with an examination of feces of a sample of school children to determine probable infection, followed by a population census of all inhabitants and their residences, with subsequent medication of *each* person, only in high-infection areas. Careful records were kept of who was treated and when. Some areas were untreated if incidence was found to be low.

A parallel program of counting of infected snails in the water of the area plus eradication of the snail through pesticides and earth moving was required. And another for improved sanitation was carried out by separate teams. Coordination of these teams is difficult since they become physically separated, and they meet unexpected delays. In addition, school teachers were enlisted to give special programs during class time; children made campaign posters and carried the message of sanitation and need to take medication back to their families.

Each of the areas was under a team leader who checked on progress and made certain that all inhabitants had been reached— unless they moved away in the interim. Once an area was covered, the team moved on to another village in the district for which it was responsible. In this way, the entire northeast of the nation was expected to be covered in several years, reaching down into the capital, Brasilia, itself.

FUTURE COOPERATION
AND ASSISTANCE

The problems facing those who seek to eradicate or reduce the presence of the six tropical diseases are numerous, complex, and arise from quite different sources—medical, political, economic, and bureaucratic.

To focus attention on the many facets of the problems in tropical diseases, distinctions need to be made among the medical aspects, the political, and the economic. Among medical priorities, the six special diseases are not high. Other diseases are larger killers and in more countries. The six have proven intractable, but not in the area of medical research so much as in the delivery process. Among political priorities, the six have been given special attention and can elicit emotional responses supporting extended assistance. From the economic standpoint, emphasis on these six depends on the willingness of governments to reallocate budgets and resources so as to generate delivery systems and markets that will attract company responses. An objective of future dialogues should be to keep all three in sight and to attempt to bring these views closer together so that similar priorities exist in all three aspects.

In meeting the problems of the special tropical diseases, several stages of activity will need to be engaged in—some in sequence and some simultaneously. The complexities and variations of these stages demonstrate the weaving together of medical, political, economic, and administrative elements at the industrial, national, and international levels.

The sequence of multiple stages is as follows:

1. Drug development is the first stage in the sense that there must be an adequate drug before an attempt can be made at elimination of the disease in the infected person. Development begins with discovery of a compound or agent that is effective against the disease. This agent must then be developed through a process of determining proper formulations and dosages plus toxicity. Success in these areas is followed by clinical trials for testing in humans as to safety and efficacy. These tests are best done on location—that is, where the disease exists so as to have proper environmental conditions as well. If success is achieved and regulatory bodies approve, the next step is a scale-up of the production process to commercial volumes. Simultaneously with other steps, but most importantly, comes the effort to commercialize the drug—that is, place it in its market niche. If all goes well, the drug will have been patented so that it can obtain a price that will repay the extensive costs of development. Finally, distribution channels must be determined and activated.

   All of the preceding activities are largely the responsibility of the pharmaceutical companies, except for approvals of the results of the clinical trials, which lead to registration with the governmental authorities. In case of tropical diseases, the companies do not always have facilities in situ for clinical trials, and they will need substantial assistance from persons on site or from institutions such as WHO.

2. Field trials are also required—taking the drug out into the country where disease exists and applying it on a fairly large scale in appropriate environmental and social conditions. The

purpose of these tests is to determine the feasibility of a large-scale eradication effort. Before governments will undertake a program of eradication, they need a fair degree of certainty that the program can be implemented successfully.

These trials are seldom the full responsibility of the companies and they are more expensive than the companies will undertake on their own. Field trials will therefore require cooperation between the companies and national governments as well as WHO.

3. Adoption of special programs of eradication by national governments is the next stage, after successful field trials. Approval of such programs will generally occur only after consideration of costs and benefits to the country as well as the socioeconomic impacts of the eradication of the disease or its continuation. This step is probably the most critical of all since without it nothing further will be done; yet no outside influence is likely to sway the national decision.

The responsibility for this stage is almost wholly that of national governments, though the World Bank and UNDP can provide funding for some of the studies necessary or even perform the studies themselves so as to provide adequate information on which policies and programs can be established.

4. An effective delivery system is required in order to implement an eradication program. This will include the infrastructure of public health services, training of public health officers as well as governmental administrators, teachers, and supporting personnel, and a careful examination of the cultural aspects of the populace so as to make sure that the drugs are taken as prescribed.

This stage is almost wholly the responsibility of the national government, and the obstacles are greater where public health services do not already exist. Where they do exist, the additional costs are for training and direct implementation plus the necessary drugs. Even these costs are expensive, as seen in the Brazilian program on schistosomiasis.

5. Vector control is necessary in most of the tropical diseases so as to prevent reinfection. This requires application of different types of chemicals (pesticides or mulluscicides) and efforts to improve the environment to reduce the locations where the vectors can survive.

The responsibility for this stage is largely that of national government; however, assistance can be obtained from WHO and possibly UNDP, and sometimes from the chemical companies also. It will need to be a cooperative effort in most instances.

6. Environmental conditions will have to be changed in order to improve general health and reduce the likelihood of infestation. General sanitation will be required as well as alterations in water supply and use by the community.

    Although this stage is also largely the responsibility of the community and national government, some assistance can be obtained from WHO and other international agencies.

7. Education of the community will be required in several directions: the proper use of the drug and the reasons for taking it; vector control and the significance of continued attention to this control; information as to the reasons for environmental changes and the methods of accomplishing it.

    Education is also the responsibility of the national government and the local community; help can often be obtained from the companies in providing educational materials.

8. Local facilities for training and research and development in tropical diseases will eventually be required in order to assist developing countries in making appropriate judgments on methods of control in education and the desirability of doing so. These facilities will provide local capabilities in decisionmaking—which were seen to be important in the Brazilian schistosomiasis program. And they will eventually support the move toward local production of pharmaceuticals that many countries insist on.

    Under its industrial development program, UNIDO has established a program for the promotion of the pharmaceutical industry in LDCs, helping them to produce domestically "the major requirements" of the country for drugs, reduce the price of intermediate materials, facilitate the transfer of technology for advanced countries, establish regional pharmaceutical industry development centers, and develop new forms of international cooperation through bilateral relations of companies and developing countries. The specific areas of cooperation delineated include the following: (a) training of pharmacists and other technical personnel; (b) establishment of adequate quality-control systems and good pharmaceutical manufacturing practice; (c) coordination of research efforts for new drugs needed for diseases prevalent in developing countries; and (d) wider use of multipurpose plants in developing countries. The pharmaceutical industry has stated its willingness to help national governments and regional centers through bilateral programs related to training, quality control, and technology transfer. But it does not consider that the suggestion for the establishment of multipurpose plants is technically feasible or economically viable, especially where medical infrastructures do not exist.

*   *   *

These stages indicate that a "total program" is involved, requiring the cooperation of several different factors. Companies have only a limited role to play. This role is primarily in drug development, but even this area is circumscribed significantly, as emphasized by the criteria that WHO sees as necessary for successful development of drugs and vaccines to be used in eradication campaigns. The agency emphasizes that these drugs must be produced at a cost that can be borne by developing countries, that their distribution and use must require minimal skills and nonspecialized supervision, and their distribution must be feasible through the public health services of developing countries. These conditions place some very serious constraints on the companies in drug development. If they are not able to comply, then additional assistance will be required in developing the market for the drugs that can be developed.

Fortunately, the WHO has begun to realize the limitations of the companies, and both the companies and WHO have recognized that cooperative programs will be necessary to resolve the problems in the area of tropical diseases.

In its *Annual Report 1977,* WHO recognized the limitations on the companies' roles and the need for expanded assistance to create demand-pull for additional research:

The increase in development costs of all drugs and the uncertainty of return on investment for drugs to control tropical diseases continues to limit research investment by the pharmaceutical industry in this area. Return on investment is determined by the amortization against a potential market at a stable price for a specific time period. The costs of all aspects of drug development have increased, but expanded demands for efficacy and safety testing and the expense of maintaining permanent screening facilities for compounds of potential interest are the major reasons for the low rate of introduction of new products to control the tropical diseases. The development of drugs for tropical diseases has also been seriously hampered by difficulties in securing adequate facilities and competent clinical investigators in endemic countries. Thus many factors add to the complexity of the task required for the research and development of these drugs requiring a disproportionately high expenditure of resources in relation to the expected return on investment. Where potential benefits are more social than economic in nature, the needs for support with public funds and effective forms of collaboration are required to ensure the availability of unprofitable drugs.[1]

WHO has determined that industry's role "lies in the development of drugs and vaccines that are not prohibitive from the point of view of return on investment." (*Annual Report 1977*). To facilitate cooperation with industry, WHO has included industry-affiliated scientists on the Scientific Working Groups set up for each of the six diseases. Seventeen such representatives had been appointed, and others were to be added as the groups expanded their activities. An industry official is chairman of the 18-man Technical Review Group advising on the entire Special Program.

Besides this participation, industry is looked to for screening of agents, for technical services under contract with WHO, for clinical evaluation of new drugs and vaccines under WHO auspices; and for training of scientists and technicians. Contracts for technical services include drug development. WHO expects to retain a "royalty-free, nonexclusive" license, to permit passing the rights to production to public bodies. This requirement will be modified only if "stronger financial incentive" is needed to stimulate commercialization of a new agent.

However, one of the strongest incentives that could be given to the transfer of technology and development of appropriate drugs would be the elimination of disincentives to industry through restrictions on industrial property rights, price ceilings that have not risen with inflation, and discrimination against foreign investors. Efforts to promote local production have often resulted in poor quality of drugs and therefore ones that are ineffective or have undesirable side effects. Yet there is strong pressure from national governments to develop domestic pharmaceutical capabilities. Accommodation will be necessary between governments and the international companies, and assistance of international agencies can be useful in bringing about effective cooperation.

In achieving this more cooperative stance, new institutional arrangements are likely to be necessary, including a variety of bilateral arrangements, between companies and independent or government-sponsored laboratories in developing countries, joint ventures between foreign and local companies, centers for training technicians and pharmacists, and regional centers dedicated to research on specific diseases.

To assist developing countries in strengthening their own research training activities, WHO has established a Research Strengthening Group, which held its first meeting in October 1977. It adopted priorities that included assistance to research and training institutions that are to be set up in tropical countries, support of training of technicians from tropical countries; assistance in the diffusion, interpretation, and integration of new knowledge in tropical diseases; and facilitation of more rapid transfer of knowledge, technology, and skills to developing countries.[2]

With the interest of the World Bank, U.N. development program, and UNIDO added to that of the WHO, sufficient international resources could be brought to bear to attack the problems of tropical disease successfully—but only if the national governments provide budgetary priorities and administrative support, and only if the interest of the companies is maintained through the cooperative approaches adopted. The problems of the six diseases cannot be laid at the doorstep of any single actor, and they will not be resolved without extensive and intensive cooperative efforts. This is clearly one set of problems in which the adversarial relations between business and government will have to be set aside in favor of a posture that puts the problem on the other side of the table to be faced collectively by all concerned.

NOTES

1. WHO, TDR/AR(1)77.14.
2. WHO, *Newsletter,* No. 10, January 1978, TDR/NR/781.

Chapter 3

Industry Ties to Science and Technology Policies in Developing Countries

For the contributions of R&D activities of international companies to be effective, they must be injected into an S&T structure in developing countries that can make use of the programs and facilities set up. Both for this reason and that of the desire to become more self-reliant in science and technology, the developing countries have urged the adoption of policies to strengthen S&T capabilities and requested the assistance of advanced countries to this end. These objectives are a major purpose of the Action Program adopted by the U.N. Conference on Science and Technology in Development at Vienna in 1979.

This study of the S&T policies and programs had as its objective the determination of ways in which developing countries needed strengthening—where they were in the process of developing science and technology, what problems they faced in further advances, what needs could be met with the help of advanced countries, and what roles the international companies might play. This section provides both a summary and a set of conclusions derived from the research on S&T developments within nine LDCs (Brazil, Egypt, India, Indonesia, Iran, Korea, Malaysia, Mexico, and the Philip-

The study on which this summary is based was written by J. N. Behrman, and published under the same title, by Oelgeschlager, Gunn & Hain (Cambridge, Mass., 1980).

pines) during 1978–79;[1] no attempt was made to characterize fully what any one country was doing—only to obtain illustrations of programs and policies. This part summarizes the issues covered in the investigations, the situation as to S&T policies, the needs for the future, and potential contributions of TNCs. Out of this study and the prior ones on R&D activities of international companies would come some recommendations as to policies that the U.S. government might adopt in assistance S&T strengthening abroad—as suggested in Chapter 4.

The formation of S&T policies in developing countries begins first with the determination of economic and social objectives of the country that S&T is to serve or to help achieve. These objectives are not simply national in scope but are subnational, having to do with regions or even villages within the country; and they are international, having to do with regional roles of the country as well as its place in the international economic order.

Once these goals are set, the policymakers can turn to the selection of technologies by which to pursue these goals. The sources of technology available are those from the international community and those from within the country. The process of selection of the technologies is itself significant in determining what technologies are chosen since groups will have different perceptions as to the usefulness of various technologies. The selection will also differ according to the users of the technology since some are more and some less capable of assimilating the technology. And some sectors can be leaders in adopting new technologies while others are followers. In selecting and developing technologies, the role of government R&D institutes is critical; they frequently have to substitute for private initiatives in developing S&T, and they are the only governmental groups capable of assessing foreign technology if it is to be screened. To be effective, however, in expanding indigenous science and technology and making it more applicable to national needs, extension of the results into the user communities is necessary, requiring some type of extension service.

The U.S. government and transnational corporations can be helpful in developing national responses in each of the activities indicated, but it is unlikely that an overall policy can be established that meets the needs of each developing country. These countries are quite varied in their developmental objectives and their capabilities of selecting or assimilating technologies. Many lack S&T infrastructures, and each is in a different stage of progress in creating its own S&T capabilities.

THE ISSUES

Two sets of issues face the United States in forming its policy toward the use of S&T in development objectives of LDCs: The first is an appreciation of the problems faced within the LDCs themselves, and the second is determination of its own posture and potential contributions to helping resolve those problems. Seven different problems can be identified that face the developing countries themselves: the developmental strategy, the structure of industrial development, the sources and selection of technologies, the generation of local S&T capability, creation of ties to technology users, control over foreign technology inflows, and generation of indigenous technology outflows. Facing the United States itself are the questions of what form of dialogue should exist between it and the developing countries on these issues, and second, what potential contributions the government or private companies can make to these problems.

Development Strategies

Pronouncements on developmental strategies of LDCs signal clearly that they are directed at the creation of national self-reliance and improved quality of life for their citizens. Self-reliance does not mean autarky or autonomy but the ability to make their own choices, including expanding the number of choices. It makes a difference, of course, who exercises the choice, both from the standpoint of development strategy as well as the role of S&T. The results will be different as between decisionmakers within government ministries or government research institutes, university researchers, or industry R&D officials. The criteria of choice applied by each is likely to be different, and the ability of each to absorb the results of their choices and make them useful for national goals will certainly differ. Also the knowledge of each as to available technologies will differ. Consequently, what self-reliance means in practice cannot be stated generally but must be identified for each country separately. In the main, the decisions as to developmental strategy will be made by governments through various institutional structures. In developing countries, these decisions are not going to be left to the market nor to private industrial companies, save for some implementation.

Each country will decide differently as to the various choices that must be made in the formation of a development strategy. These choices involve the relative emphasis on agriculture versus industry, the priorities among different industry sectors, the roles of the public versus the private sector, the objective of satisfying the demands and the needs of the masses versus those at the top of the social structure, priority to the domestic versus export market, and finally, the specific priority to be given the role of S&T in implementing these choices. All these choices require an exercise in political will, and the basic problem in the area of S&T is that governments have been slow to identify S&T as a critical input in the development process—yet it is equally or more important than capital was seen to be in the 1960s.

Without a developmental strategy, it is not possible to formulate policies toward S&T. Without it, the use of particular technologies will simply fall out of a set of uncoordinated decisions by various actors. This has happened for a number of decades, and the present emphasis on S&T policies is an attempt to rectify that situation. Once the development strategy is determined, attention can then be given to S&T roles in specific sectors.

Industrial Structures

The selection and application of technologies in pursuit of developmental goals occurs within a concept of the precise agricultural and industrial structure sought. Since this study was not concerned with agricultural research, the comments here will focus on the industrial sector. Most developing countries have followed a policy of selecting industries on the basis of import substitution rather than looking at their future place in worldwide industry and how they might achieve it. They are only recently beginning to recognize the limitations of this strategy. Decisions must now be made as to the extent to which domestic materials are processed before exported, the level and scope of agri-industry, the extent to which capital equipment is produced locally, and the emphasis on particular activities within the intermediate-goods sector, consumer durables, and nondurable consumer goods. Again, for each of these the market needs to be defined—both geographically and in terms of income levels of consumers.

Many developing countries do not have such detailed development plans, and some do not even have a method for making the determinations. Yet whoever makes these determinations is in a position to advise on the selection of technologies, though these may be left to those who make the investment decisions within the sectoral priorities, however uncoordinated or unplanned they may be.

Selection and Source of Technologies

Technologies are obtained either from domestic sources or from abroad. The selection in some countries (or sectors) is left to the direct users and in others is stimulated by or screened through governmental mechanisms. Developing countries are increasingly concerned to shift the source of technologies from the advanced countries to indigenous sources; but there are a number who recognize that foreign technologies will be required indefinitely, and the only question is which ones are obtained and under what terms.

Among the foreign technologies, it has been estimated by experts at a pre-Vienna seminar in Jamaica that about 40 percent of what developing countries need for their stated objectives can be obtained from publicly available sources with little or no cost to the recipient save that of gaining access; probably another 30 percent is in private hands but not "owned" and can be obtained at virtually the cost of a transmittal; roughly another 20 percent is proprietary information and must be paid for in amounts that reimburse not only transmission costs but also some of the costs of generation; and a final 10 percent is available from governmental institutions on varying terms. Frequently private recipients in developing countries merely take the first source available without looking in alternative sources of similar technical information.

Only the 20 percent that is proprietary raises serious questions about the costs of the technology obtained from abroad. However, all technologies raise questions about their impacts when employed in the recipient country. Cost issues relate to direct payments requested by the donor, ancillary terms in the contractural agreement affecting the rights of the licensee to utilize the technology, the preparation time on the part of the license in getting ready for the negotiation, the cost of the assimilation of the technology—all offset by the various benefits, including a variety of technical assistance and training benefits provided by the licensor. Many of the developing countries consider that these costs are too high—not necessarily because of poor valuation procedures—but more likely a result of the ignorance of the recipient in bargaining effectively. Recipients frequently do not know the impacts of the use of the technology nor alternative types or sources. Since the licensor developed the technology, he is more able to understand its contributions and problems than is the licensee. To help balance the negotiation, developing-country governments frequently accept the responsibility for determining the usefulness of the technology and the conditions under which it is appropriately transferred, even to approving specific terms of licensing agreements.

When they inject themselves into this process, some governments have sought to depackage the inflow of technologies so that a local research unit could undertake some phase of the technology creation, adaptation, or assimilation. Rather than being squeezed out of the market entirely, because of an inability to supply the entire package, this tactic is supposed to open up opportunities step by step for local suppliers of technology. Difficult problems arise of distinguishing the pieces of a package so it might be separated; of valuing them so that a cost can be applied to the remainder; of guaranteeing the results of the total package when part of it is separated out; of providing technical assistance when part of the package comes from a separate source; and of providing a continuing flow of adaptative technology when the local supplier may not be able to keep up with new technologies in his particular segment. This technique was employed to an extent in inflows of technology into Japan; but the recipients had rather advanced capabilities, knew what they needed, and were not concerned with guarantees. When these conditions did not obtain, the entire package was bought, and even continuing technology. Since this tactic appears highly attractive to developing countries, even though the underlying conditions are different than in Japan, it remains a serious issue in the determination of sources and of specific technologies.

Generation of Local S&T Capabilities

In order to achieve self-reliance and diminish dependence on foreign technology, each of the LDCs visited is attempting to accelerate the development of local capabilities in S&T. Only a few have faced all the problems squarely and only one of those visited has shown substantial success in resolving the problems. Among the questions that need to be resolved are those of the distribution between public and private facilities in the generation and transmission of technologies, reliance on ministerial or departmental institutions as compared to autonomous governmental institutions to generate R&D, the development of coordinating mechanisms (both institutional and personal), support of education and redirection of curricula so as to convert educated manpower into the required professional skills, effective communication with users to guide S&T programs, and the stimulation of R&D activities by industrial companies.

Each of the countries visited has resolved these questions differently, though some have not addressed all the questions and are therefore still formulating their positions.

Ties to Users

Technology, whether sourced abroad or home, is of little use unless it is applied in the productive process by the recipient—and effectively enough to improve the firm's competitive position and thereby move the country toward its economic and social goals. At least four different users may be identified; each having multiple purposes for the technologies employed; governmental ministries, departments, or agencies; state enterprises; private national enterprises; and affiliates of foreign companies. Each has a different set of ties to the sources of technology and a different interest in the country's developing its own S&T capabilities. Since the objectives of each differ, it is significant which are selected to be represented in the process of formation of S&T policies for the country.

Control over Technology Transfers

Governments of LDCs have increasingly adopted a policy of screening foreign technologies coming into the country to increase their appropriateness, to encourage utilization of local technologies, and to expand opportunities for local employment and use of indigenous materials. They have also sought to strengthen local R&D institutions by using them to evaluate foreign technology inflows.

An interesting switch has begun to occur in the attitudes of some of the developing countries as the result of the fact that they are themselves now purveyors of technology that they have developed. This has caused some of them to recognize that royalties must pay for unsuccessful R&D efforts in unrelated activities, that appropriation of the rewards of invention is more easily accomplished with patent coverage, and that a number of ancillary provisions in technology agreements are required to protect the validity and use of the technology. Consequently, in developing their own policies toward technology outflows, they are beginning to reassess their postures toward inflows.

North-South Dialogues

From the tone and expressions in the interviews in the nine developing countries, it became clear that a shift in the attitude of developing-country officials has occurred in the past several years. Rather than the former fearfulness of negotiating with the transnational companies and a concern over local inability to obtain satisfactory agreements, they now see themselves as much better prepared, more knowledgeable, and therefore capable of dealing with counterparts in companies or advanced country governments. They also see that they have something to offer that is attractive to TNCs, in terms of potential markets, and they are more able to assess the trade-offs required between company capabilities and objectives and national goals.

Further, though they reject a continuing technological dependence, they recognize that they will not for many years be able to stand on their own technologically. (Obviously, no country, not even the United States, is self-sufficient in S&T; but the goal is an acceptable balance of indigenous and foreign sources, not complete independence.) Therefore they recognize that it will be necessary to bargain for many years over the flow and contribution of technology coming from abroad.

Given the specific nature of technology and its uses, they are primarily concerned as to how to obtain and use it for *national* (rather than regional) purposes. To this end, almost all the officials interviewed expressed a desire for more frequent and extensive bilateral dialogues.

One of the policy issues facing the United States, therefore, relates to the need to determine the format of any dialogues that it would like to enter and the nature of such dialogues—that is, who would participate, at what levels, and on what subjects.

Contributions to LDC S&T Development

It is of little use to engage in dialogues, however, unless the advanced countries are willing to offer suggestions as to how to meet problems faced by the LDCs, especially in ways that involve mutual ccoperation. From the standpoint of the United States, it has some means of direct assistance, but, by and large, the needs of the developing countries would be more readily satisfied through resources available to the transnational enterprises, which are still regarded askance in some of the developing countries.

The U.S. government must also decide how it is to respond to initiatives from LDCs that propose more international institutional assistance and the formation of regional S&T capabilities.

*   *   *

These are the issues that surfaced in the discussions with LDC government officials; they were the subject of debate at the conference in Vienna, and since the UNCSTD Action Program recommends determination of appropriate means of strengthening S&T in developing countries, they will be the subject of any continuing dialogues to specify bilateral assistance programs. Although the situation in each of the countries is different, there are a number of facets of S&T policies that are similar in the sense that they are part of each picture, though with different intensity and impacts. Despite the fact that the situation cannot be generalized as universally descriptive, the following section provides a summary of the situation in the countries visited.

THE PRESENT SITUATION

The present situation in the countries visited is characterized according to their developmental priorities, the nature of their science community, the governmental R&D institutions, the roles of industrial companies, and the linkages among the various S&T institutions in each country.

Developmental Priorities

Each of the nine countries is in a somewhat different state of development, and only four of them have well-articulated development plans—Brazil, India, Korea, and Mexico. Even these differ as to the extent to which S&T policies have been articulated in support of the overall objectives. Korea is much more advanced than the others; India has numerous R&D institutuions, but they are not well directed toward development objectives; and Mexico has just begun to implement its policies. Malaysia is in the process of formulating very careful policies. Indonesia has assigned responsibility for formation of its policies, and the process is beginning. Egypt and the Philippines are still further behind; each has some R&D institutions, but they are not well directed toward defined developmental goals.

The Korean success and the lack of success in other countries shows that personal leadership from the top is a necessary part of the formation and implementation of S&T policies. Not only is a science leader necessary but also the chief executive must give his support in order to circumvent in-fighting among the various ministries and bureaus and to make certain that the funds are made available as necessary. In addition, this leader's help is instrumental in popularizing science as a career and in redirecting educational efforts.

Finally, progress in developing and implementing S&T policies requires a recognition that the problem is itself systemic, and requires the coordination of several institutions and development of channels of communication among them and with the ultimate users of technology. These institutions include the university system for the preparation of professionals, various governmental R&D institutions, industrial companies in the public and private sectors, and the linkages among them.

The Science Communities

In each country except Iran, long periods of colonialism have had a strong influence on the development of S&T institutions. Either the metropolitan country had built some embryonic R&D structures, or they had prevented them from being established. In the former case (India, Indonesia, and Egypt), the institutions formed the base for expanded effort on the part of the newly independent governments. In Mexico, many of the prior institutions were permitted to atrophy; while in Korea, a new institute was established to avoid having to reform existing technical schools.

In most of the countries, the educational system still bears strong imprints of the colonial period, including curricula that are similar to those in the metropolitan countries. Those formed by European countries remain highly theoretical in the scientific disciplines, with little attention to laboratory research or industrially oriented experiments.

In all the countries visited, attention is being given to the provision of scholarships, the redirection of curricula, and the establishment of research labs in universities. However, in virtually every instance, established universities are difficult to change, are manned with professors who are not skilled in laboratory research, frequently not full-time, and who write for international publications to obtain their prestige and promotions. University research is frequently budgeted along with a variety of other university activities and therefore often does not have a high priority.

Training of professionals overseas is not an adequate substitute for development of appropriate educational facilities and orientation in the country itself because the professionals frequently do not return. Even when they do, they are imbued with methods and orientations of the advanced countries, including the desire for sophisticated equipment, which frequently cannot be used effectively or maintained. Consequently, many of the graduates in the scientific disciplines do not find their way into research institutions where prestige and pay is still not equal to that in a professorial post or in an administrative position within the government.

Virtually every government recognizes the need to expand significantly the percentage of scientifically trained personnel, and for this purpose, Korea established an independent institute of advanced studies, circumventing the problem of changing existing curricula and faculties. One of the limitations on the development of governmental R&D institutions is the lack of appropriately trained scientists and technicians. Therefore many of the LDCs are turning their attention to university programs to make certain that the content of programs is appropriate for the needs of science institutions and science-based companies. This in turn requires that the professors themselves be adequately trained—meaning that they develop an orientation of industry-related, rather than purely theoretical, research. Government support is generally in the form of grants to universities for the development of particular programs, for research projects, scholarships, professorial exchanges abroad, and so on. The laboratories of various departments are also used as back-up or as fundamental research units related to the more technologically oriented governmental institutions.

The reorientation of university programs—especially those of engineering schools—is not really effective unless there is also a close tie between the universities and the industrial sector. In most of the developing countries, this link has not been formed for a variety of reasons. One is the general distrust of university orientations on the part of the private sector, with the result that many of the scientifically oriented students wish to go to work with the government rather than the private industrial sector. This is frequently the case even though governments may pay less because the longevity in the job is greater and the prestige is frequently higher. In addition, in some of the developing countries the perquisites that go along with a governmental job include the ability to moonlight in one or two additional positions, so that the income of an individual is raised substantially.

Government R&D Institutions

In view of the absence of indigenous R&D capabilities, and to start their gestation, many LDC governments have established embryonic science institutions, generally beginning with institutes for standards and measures and rudimentary facilities to encourage quality control on the part of local companies—especially for exports.[2] A few countries have established a number of national laboratories dealing with specific sectors or disciplines to provide alternative technologies to those available from abroad. How to get these technologies adopted, how to create the ties between these institutions and local industry, and how to make certain that local technology is an adequate substitute for the foreign technology—all of these are as yet unresolved problems in the creation of state-owned R&D institutions.

As a consequence of the uncertainty as to the best way of setting up S&T institutions, there is a wide variety of organizational structure and coordinating mechanisms among the countries visited. Some have relied on coordination through a specific ministry; others through independent organizations; others through an interdepartmental agency; and still others through a science unit reporting directly to the president. In each case, successful coordination depends on the authority given to the entity responsible, whether or not it is given control over firms, whether it has responsibility for program review and evaluation, and finally on the personalities of individuals in the top positions. In several countries, responsibility was assigned, but authority was not provided. For example, in Egypt a coordinating unit was made responsible for pulling the program in S&T together; but it had no authority over funds or program review. In Korea, the coordinating responsibility is with the Ministry of Science and Technology; in India it is with a National Committee on Science and Technology.

The most effective coordinating mechanisms appear to be those that are also R&D operating units themselves and not just controllers of funds. The responsible authority needs to have the respect of those R&D laboratories reporting to it, and this can be achieved more readily if it is staffed with individuals who are familiar with day-to-day problems of research.

For success in research related to industrial development, the relevant institutions need to be somewhat autonomous from the normal programs of governments so they do not get tied up in a competition for funds as well as in the political jockeying that tends to favor one laboratory or sector over another. Where autonomy was not provided, it was found that government R&D institutions were predominately concerned with S&T support of governmental programs rather than those in private industrial sectors.

The gestation of governmental R&D institutions tends to begin with the formation of a laboratory related to measures and industrial standards. This step seems to be somewhat in response to the need to meet international standards in exports of components or final products, but it is also being pressed on the countries because of consumer safety and environmental concern. However, not all countries visited have moved as far as they need in this direction. Many standards are not adopted by local companies, nor even by state-owned enterprises. Few foreign affiliates have adopted the standards since they tend to meet higher ones imposed by parent companies.

Several standards institutions have developed "marks of quality" to stimulate an improvement in quality on the part of domestic producers, but this initiative has not been supported by local industry—many of which have complained about pressures to raise quality standards. Only a few standards are mandatory in most of the countries, and much needs yet to be done in coordinating the standards adopted among the developing countries themselves.

The development of standards institutions is a necessary step in the gestation of S&T needs in developing countries, partly because the institutions are frequently a starting point for the development of applied scientists and the insemination of a technology-innovation attitude on the part of local companies. These institutions also begin embryonic ties with companies through technical assistance on quality-control procedures.

Despite the need for support of agriculture and agri-industry based on local resources, few of the countries have significant R&D institutions working in this sector. It is undoubtedly a difficult sector to assist in technical innovation, and some significant efforts are being made in the Philippines on the use of local wood products, attempting to promote new product uses, and in Malaysia on the use of rubber tree wood, which requires fairly sophisticated research technologies because of the peculiar qualities of that tree. The difficulties in agri-industry R, D, and E are illustrated by the slow growth in the relatively small engineering program within the International Rice Research Institute (IRRI) located in the Philippines. It was begun some thirteen years ago, with a grant from the Ford Foundation, to develop machinery for rice cultivation; it attempted to do some initial research on its own, to no avail. It later shifted to modifications of existing machines, and succeeded after six years of effort in introducing some rudimentary machines usable in rice plots in Southeast Asia. It eventually developed a number of machines able to till, transfer seedlings, and harvest—under manufacturing processes that permitted sale of the machines at around

$400 each. The costs of RD&E amounted over the 13-year period from 1965–1978 to around 5 percent of sales value of all manufactures throughout some ten countries. This low percent of sales was achieved only after the relatively large volumes of the last couple of years were achieved. If manufacturing companies had been forced to pay the R&D costs in the initial years, they would have paid 10 to 15 percent royalties. No single country market could have supported the necessary RD&E out of company revenues. Nor would any government have seen the market as significant enough to warrant its support of such an activity.

Therefore agri-industry research is justifiable in terms of government priorities only where the agriculture activity is a significant part of the national income such as rubber growing and processing in Malaysia. Otherwise, it appears that many of the countries are too small to support the necessary agri-industry research, which is better done by international cooperation such as through the IRRI. One field in which much more needs to be done is that in food processing and preservation. This is a fertile area for RD&E in virtually all of the countries visited, especially as it relates to general nutrition of the population.

Industry-related institutions that are tied to state-owned companies are the most adequately supported and tightly integrated of the various R&D units in these developing countries. For example, PEMEX's petrochemical research is the largest single budget item in the Mexican government's expenditures in the manufacturing area. All other industry-sector research sponsored by the Mexican government is selecting a few industry sectors on which to concentrate—one of them being aircraft, along with petrochemicals. The Egyptian state R&D institutions are closely tied to the government departments and the companies they run since very little private industry remains.

Although India has the largest number of governmental R&D institutions, they are little related to industrial development, except where state enterprises are involved. A few laboratories are specifically focused on technologies useful at various levels of industrial activity—for example, the village or cottage level, subregional levels in the country, and for export.

Few governmental R&D institutions are autonomous, as they are in Korea; most are tied to a ministry and its budgets as well as its program objectives. For example, health research in Egypt is tied tightly to the Ministry of Health, which is responsible for the state-owned pharmaceutical companies. Malaysian R&D institutes are tied to the Ministry of Science, and the Department of Science and

Technology in India has overall responsibility for the operations of a large number of R&D laboratories, including that oriented toward industrial support.

The most developed autonomous structure is that in Korea, which was instituted some years ago under a grant from the United States. The formation of the Korean Institute of Science and Technology has been the source of a gradual spin-off of several industrially related institutes, including an Advanced Institute of Science, a technology assistance group, and an information service. Out of these various institutes comes a set of extension services reaching into the local companies.

The Philippines illustrates a different approach, with the two major industrially related institutes having been set up at the instigation of the industry associations but with governmental funding. Indonesia has established an autonomous but hardly centralized authority over R&D activities with funding for the various labs coming through the central unit. Mexico has a centralized responsibility, but a number of competing units seek R&D funds for the government so that the unit principally responsible does not handle even a majority of the funds for S&T in the country.

Given the competing claims for S&T resources, the facilities available to any given unit in a country tend to depend on the political influence of the lab director or the internal politics within the coordinating mechanism. In some instances, industry support is critical; or "national or departmental needs" are translated into R&D programs, without regard to the longer term needs of industrial development.

The activities of the units doing industrial research are varied. Few countries have designed the same types of programs, and they are not all done equally well. Resources are not readily available, and stringent priorities have to be set. Consequently, the major activity for governmental R&D institutions is in direct support of governmentally budgeted programs that are the responsibility of departments or ministries or are undertaken by state enterprises. Given the fact that the industrial sectors in many of these countries are either predominately state owned or are composed of medium- and small-size companies that are not yet technically oriented, there is little pull from the industrial sectors on R&D capabilities in these countries. Korea is a marked exception. It has proven that contract research can be undertaken by governmental R&D institutions and that this can be a major source of funding and means of assuring application of results. However, it is not clear that the Korean experience is transferrable completely into any other setting because the same initiative and receptivity on the part of private entrepreneurs does not exist in other countries.

Apart from response to industry needs or innovations, R&D institutions have included within their programs the training of workers in companies as well as training of government technicians and engineers. Several of them also provide courses for education of workers entering the labor force or seeking new jobs or skills.

Within the programs of R&D institutes will be found also a variety of technical assistance projects seeking to meet or anticipate problems faced by their particular clientele, including the inculcation of quality-control procedures. Part of this technical assistance eventually develops into an extension service or the promotion of inventions. And this last may be institutionalized as in the National Research Development Council of India which takes innovations from throughout the country and attempts to merchandise them to potential users.

The various institutes also include within their program a variety of test facilities, including testing of imported components or final products, commodities for export, quality-control instruments, and manufacturing equipment as well as test equipment in manufacturing plants. Some of them also provide facilities for repair and maintenance, especially for medium- and small-size companies that do not have the ability to purchase such equipment.

Finally, programs of these institutes often include educational support projects such as the provision of facilities or research by Ph.D. candidates and the seconding of researchers into positions as adjunct professors in engineering and technical institutes.

In no country was it found that a lab director was wholly satisfied with the developments to date, and the general needs were for more training and better research personnel. For example, many of the research units are quite small (two to four researchers) with budgets eaten up by wages and salaries, meaning that little remained for experimental equipment. Even where equipment had been obtained, it was frequently left unused because of incompatibility with equipment donated through international cooperation arrangements or because of breakdowns and the absence of facilities for maintenance and repair, as well as the lack of exchange for purchasing spare parts. Frequently national budgets did not provide for adequate maintenance because of underestimating this part of the cost of R&D activities.

Company Roles

In virtually no case among the countries visited was local industry significantly involved in R&D activities. In Indonesia, Iran, Malaysia, and the Philippines, companies are far from this stage of development, and in the rest only a few of the largest have begun embryonic R&D laboratories. India has a few companies with substantial R&D labs and both India and Mexico are beginning to export some indigenous technologies. By far the predominate pattern, however, is for the industrial sector to find its technologies abroad and to bring in tried and proven technologies for application, even if the result is that the user is restricted to a defined market. Technology from foreign companies provides the fastest commercialization possible and reduces the risk-exposure of recipient companies. Even state-owned enterprises were reported as leaning toward foreign technologies, despite the existence of potentially useful domestic technologies—sometimes developed within the same ministry. Consequently, some of the governments have attached R&D facilities to the state-owned companies so as to force a shift in the source of technology from abroad to a local entity. Where this has been the case, the generation of a new institute has faced several obstacles, not the least of which is the desire of scientists for administrative or "white-coat" positions, rather than those jobs having to do with plant engineering or on-site processes, or pilot projects, which would be necessary in scaling up laboratory models of a manufacturing process or a product.

Governmental screening, evaluation, and registry agencies have been set up in each of the countries to control the inflow of foreign technologies and to make certain that what is obtained is of the greatest use for the pursuit of developmental objectives and at reasonable costs and terms.

R&D activities of affiliates of transnational corporations are also embryonic, with the largest labs in Brazil and India, where the markets are largest and are increasingly sophisticated. These initiatives start with quality-control and technical-service activities, gradually moving to materials testing for substitution and adaptation of local materials, as foreign exchange costs rise, and eventually to product adaptation as the market grows. By and large, however, these affiliates also rely on foreign technologies—almost wholly from the parent—and this reliance is not changed when the affiliate is a joint venture with a local partner, unless of course the local partner already had an R&D unit and the tie-up was partly for the purpose of utilizing these capabilities. Such a situation might arise in India with a few of the companies there and will arise in the future in Brazil and Korea.

A few of the governments have begun to consider how to stimulate R&D activities of these foreign affiliates, looking to the foreign

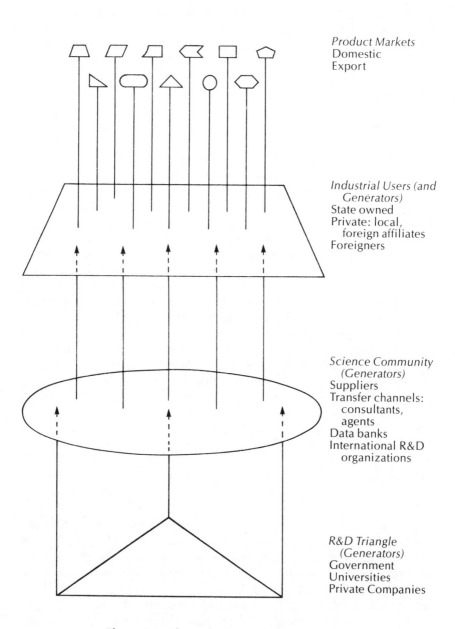

Figure 3.1. Flow of science and technology.

investment regulations or other incentives to induce the introduction of initial and later stages of R&D. The objective is not only to provide greater employment for scientific and technical personnel, thereby stimulating careers in science and technology, but also to accelerate the process of diffusion of technology throughout the supplier-customer system and to stimulate a demand for technologies on the part of local industry, it is hoped from domestic R&D institutions.

Linkages

The existence of R&D activities in universities, government institutions, and the industry sectors is by no means enough to develop S&T capabilities in the developing countries. Of most critical importance is the linkages among these various units.

The literature on development of scientific infrastructure has presented the concept of a triangle forming the base of a nation's scientific and technological efforts. This triangle is composed of the *government, universities and independent institutes,* and *industry.*[3] All these are generators of R&D in some way, but each also has a special and distinct role. The educational institutions are to supply capable manpower; industry is to supply the opportunity to utilize scientific and technological results; and government is to stimulate and guide these efforts toward national goals. From this study of government S&T policy, it is clear that at least three other elements are needed: The fourth is a *science community* oriented to industrial research coming out of the three previously mentioned groups; it will not arise unless encouraged to do so. It is broader than the basic triangle in that other members comprise it such as consultants, intermediaries or transfer channels, data banks, and all are grounded in a popular orientation to science and technology. The fifth is a body of industrial *users,* who apply the technology in cost-reducing processes or in new products. And sixth is a *market* that elicits and utilizes S&T results, making the effort profitable. Appropriate market pulls (domestic and foreign) will not always exist even if the triangle of research and technology operates properly. But the triangle will not be successful in generating industrial progress without sufficient market pulls.

These interwoven elements determine the capability of any given country to carry out an effective S&T policy, to receive technology transfers effectively, and to stimulate indigenous R&D efforts (Figure 3.1). Each plays different and multiple roles, though all are linked both horizontally and vertically. Thus both governments and companies are found in each of the four stages, from generator to market. They are linked also in the process of (a) determining appropriate technology for defined economic and social goals; (b) discovering the availability of technology (foreign or domestic) and how to transfer and diffuse it; (c) generating technology locally and translating it into commercialized goods and services; and (d) tying affiliates of international companies into domestic S&T policies.

Determination of Appropriate Technology. The government holds the ultimate responsibility for the assessment of the appropriateness of technology in reaching the economic and social goals of the nation. But this ultimate responsibility does not necessarily translate into *specific* responsibility for assessment of each and every technology applied. The government may seek such a total responsibility, but it may also reject it, preferring merely to set the rules by which technology is applied and to support development of particular types of technology for selected purposes. The difference depends on the extent to which decisions of this sort are left in the private sector or brought into state agencies. Therefore a critical decision for determination of appropriate technology is the scope and nature of the private sector compared to the public sector in industry.

Even the procedures to be applied in the assessment of technology are the ultimate responsibility of government, in the sense that legislative rules determine who is to make the decision: independent buyers and sellers, approved private institutions, professional associations, government licensing agencies, regulatory bodies, or legislators. But each of these groups will differ as to their ability to assess technology—not only in terms of meeting national goals but also in terms of understanding its impact on the total environment.

The first set of linkages therefore is between the government and the science community. Although the government has basic responsibility, it cannot perform its function without an active science community that can contribute to (or check) the judgments of the various units involved in the assessment process; the educational institutions must have been able to provide appropriate skilled individuals; and private industry will have a role to play in the assessment of what is appropriate in terms of the signals that it obeys.

To develop linkages that do *not* provide for technology assessment—whatever the criteria or processes—is simply not to have pushed S&T policy to its ultimate purpose of achieving contributions to economic development greater than the costs of innovation and commercialization.

Availability of Technology. The establishment of processes for the determination of the appropriateness of technology would be vacuous unless there were a flow of technology into the system that could be assessed for its contribution. Therefore technology must be available. It is available from two sources—from within the country or from outside. (It is not always the case that technology is available from within the developing country, though it is an exceptional country that generates *no* technology of any sort.)

The development of local capabilities for the generation of technology obviously should be directed toward the goals of development, *unless* the objective of science is merely to assuage the curiosity of scientists. In fact, in many developing countries, indigenous sources of technology are not always directed to the needs of the national economy, as shown earlier; and even when they are, the effect is inadequate for lack of resources or for lack of appropriate orientation to industrial needs or demands. Development of a scientific infrastructure of this sort demonstrates some science capability, but it does not necessarily add to economic development of the country itself. Industrially oriented research arises only when continuing *links* are formed between the industrial *users* of technology and the *suppliers*.

To form these linkages requires an input from the government, as well as from the industrial sector, into the universities. And this input process needs to be institutionalized through advisory groups.

To tap the technology that is available from outside the country, the first requisite is that knowledge exist of what is available and from where. Here there is a large gap in the developing countries. The second requisite is that there be knowledge of its usefulness and the terms and conditions under which the technology is available— to permit a cost/benefit assessment of its contribution to economic and social goals. To have this knowledge requires not only that there *be* sources of information abroad but also that the developing country have within it a scientific community that is in touch with these sources of information and can understand what is being provided them—i.e., that it have *links* to the outside sources. An educational and scientific infrastructure is needed that is linked into the needs and goals expressed by the government, the external sources of information, and the channels through which the technology would be transferred, mainly the industrial sector.

This knowledge of the technical availabilities must be complemented with an understanding of the potential terms and conditions of the transfer including the channels through which the technology can be effectively transferred. That is, who can be an effective recipient—government, universities, independent institutions, local companies, foreign-owned affiliates, or one of numerous types of consultants. Continuing links with these channels of transfer permits assessment of their different capabilities in making effective transfers, facilitating assimilation, and negotiating acceptable terms of transfer.

The terms and conditions of technology transfer vary widely, depending on the type of technology, the length of its use, the ownership conditions, and so on. Over thirty different provisions are frequently included in licensing agreements, any of which can and will be modified according to the terms in the others. A complex negotiation is normally required, and negotiators should be knowledgeable about alternatives; otherwise they may not be satisfied, even if the terms they obtain happen to be quite appropriate—that is, they may not know how well they succeeded.

Local Generation of Technology.   Five independent sources of potential technology generation exist within a country, each of which is most effectively utilized if they are linked with some of the others. They include governmental ministries, agencies, and institutions related to science and technology; independent research institutions and consultants on technology application; university schools and laboratories; indigenous companies; and affiliates of international companies. The scope of operations, orientations, capabilities, and contributions of each of these is different, but they can and should be mutually supporting at least at the margins. And some of them cannot live and grow effectively without the support of one or two of the others. In fact, a complete structure of technology generation in a country requires government inputs, a university contribution, and industrial use. With these the basic research triangle can be formed, creating the necessary condition for technology generation and application. The industrial corner of the triangle can be filled by affiliates of international companies, even if local companies do no research and very little application. But it is obviously better if indigenous companies are also engaged in the same activities.

If this basic triangle has been developed, there is a possibility of gradual diffusion, which forms a science community and elicits responses to the market, accelerating the generation of local technology. Local institutes (e.g., for medical research in hospitals or engineering consulting groups) and local companies in each industry sector will be stimulated by the existence of these three elements in the scientific and technical infrastructure. For example, pharmaceutical companies need clinical research and testing to support the investigation of efficacy of particular drugs and dosages or formulations; construction companies do not have adequate research and test facilities within their own capabilities and frequently rely on engineering consultants; and local companies will begin to see the desirability of adopting particular technologies and seek to apply those used by competitors, frequently hiring away technicians so as to begin their own quality-control or technical-services capabilities and embark on the road toward developmental research. This diffusion occurs in some countries through mobility of scientists and technicians among different institutions. However, in some developing countries this mobility is restricted because of the lack of jobs providing sufficiently high income and prestige. On the other hand, multiple job holding at the professional level provides a diffusion that otherwise would not exist as a result of the lack of mobility. This multiple job holding provides an opportunity for mobility of ideas and activities without the actual severance of ties that job hopping would produce in another setting.

Market Pull. The preceding steps are incomplete in describing the role of science and technology in industrial development until they are complemented by the development of a market that is sufficiently demanding, sophisticated, or pervasive to pull the available technology through the system, leading to the development of cost-reducing manufacturing processes and market-satisfying products. The generation of R&D capabilities in electronics communication, mass transport, medical and health research, research on foods, cleaning products, rubber products, and chemicals of a sophisticated sort is not likely to occur until there is a perceived demand for more rapid and extensive communication facilities.

The diversity of demand and the level of its sophistication will determine the diversity and sophistication of research activities. This does not mean that they will match precisely because it still may not pay a given company or government agency to go into the more sophisticated levels of research if the results can be more readily obtained from abroad. Its own effort can then be concentrated on dissemination of existing technologies at the lower and middle levels of technology.

The gradual development of the market in an LDC is likely to produce the following sequence of growth in an R&D activity, especially in a private enterprise:

The introduction of quality-control capabilities so as to make certain that the product does in fact meet the standards required by governments, the market, or simply good practices.

The introduction of technical services capabilities to supply the customer with continuing advice on the use of processes, materials, components, and the servicing of the final product.

The introduction of facilities to test local materials for substitution for previously imported product or materials.

Establishment of facilities to adapt foreign products themselves to local needs, especially a mass market of low-income persons.

The addition of applied research capabilities to permit translation of foreign-generated basic research into developmental activities, including product design.

The introduction of pure research capabilities—though this may not occur for some decades in many of the presently developing countries; in fact, there are arguments for not introducing this capability in a number of countries since it is more efficient and effective to do pure research in a few centers and spread the results around the world.

The precise structure, extent, and growth of the market in each home country will determine the usefulness of these different capabilities in pursuing developmental goals. However, it must be stressed that the market in many of the developing countries is made up strongly of governmental demands—as in the health field, for example. The government therefore is both a provider of research capabilities and a user of the technologies since it also often constitutes the market (see Chapter 3). In this instance, the government can provide all necessary links since it can decide the demands to be met and the extent to which it will help meet them through domestic generation of R&D capabilities.

* * *

In no country was an adequate set of linkages found to exist. The major gap in all is the pull of the market *through* the initiative of local companies. This is largely because of the lack of development of these companies in their appreciation of the usefulness of technical innovation but also the ready availability of international technology to serve what appear to be the market demands—which may or may not be the needs as determined by the government. Therefore there is a set of divergent signals: potentially weak ones coming through the market that are most readily satisfied by existing technology rather than more costly locally developed technology, and another coming from governmental programs that may require local R&D activities. Even here state enterprises are frequently found to turn to international technology rather than wait for locally developed innovations. Only in Korea did it appear that significant efforts were being made to form this particular link between the private sector and R&D institutions supported by the government. Others recognized this gap and considered that it needed a higher priority but seemed to find the techniques of formation too difficult to introduce.

The strongest link is between the universities and the government institutes since governmental institutions are a source of prestige employment for scientists, providing also some degree of security for researchers. Therefore attention is also given by the government to the development of university programs in research, though not necessarily oriented to industrial needs.

The university link to the industrial sector is also weak since the industrial sectors in most countries do not respect the scientific ability of university researchers, which they consider not to be oriented toward industrial problems. Even when industrial representatives are requested to assist in advisory councils for the development of university programs, we find few such representatives in place, and in some instances a withdrawal.

A few of the countries—Malaysia, Iran, and Egypt—have recognized that they are not yet at the stage of industrial development that would call forth the required linkages, and therefore they are not greatly concerned about their absence at present. However, it is clear that even where more advanced industrial development has occurred—as in Mexico and Brazil—it is still hard to establish these links, and a high priority needs to be assigned to the task as well as appropriate resources dedicated.

NEEDS FOR DEVELOPMENT OF
S&T PROGRAMS

From an assessment of the elements making up S&T policies and an examination of where many of the countries are in their present situation, it is clear that there are some substantial needs that remain to be filled if S&T is to play an effective role in meeting development objectives. Again it must be remembered that developmental objectives are not the same. For example, China, Japan, Korea, and Taiwan all wanted to industrialize in ways that would directly help the poor to elevate their standards of living. There are other countries that have had less concern for the bottom levels of the society and are seeking to pull the society up from the top—as has been the case in India and in some of the Latin American countries.

To direct the use of S&T to the removal of poverty, the top governmental leaders must direct and mobilize S&T to that objective. It appears from the analysis so far that even having R&D institutions in place will not produce the desired results in the absence of strong leadership from the top. Where the top leadership is fearful of alleviating the poverty of the masses because of potential opposition from the people as they rise, little can be expected in fostering S&T policies. Frequently what is discerned in the abstract as an "obstacle" to the use of S&T in development turns out to have the physical body of a "landowner," a "general," a "president," or "legislator" who objects to mobilization of resources for the poor or to placing a high priority on meeting their particular socioeconomic problems.

A second unfilled need in most of the countries is an adequate educational infrastructure—beginning with the development of curiosity in a family setting and going through the postgraduate disciplines. Many countries lack materials at the secondary levels of education to stimulate scientific inquiry, and most present teachers could not use them effectively without further training themselves. The substitute of sending students to the advanced countries for training is not cost effective and frequently produces scientists with an inappropriate orientation for work in their own countries. Rather than trying to adjust curricula in the advanced countries, a concerted effort is needed to reorient curricula in the developing countries themselves toward industrially related research. Much more advance in the developing countries is likely to occur from the development part of R&D than the research part, which requires courses oriented toward innovation, adaptation, modification, and commercialization. For example, more can be done in attention to dosages and delivery systems for existing pharmaceuticals than will likely be accomplished by a new drug breakthrough against any particular disease. To achieve these reorientations, more attention needs to be given to the technical schools in all disciplines, including the process of equipment maintenance and repair.

A third fundamental need is the creation of *autonomous* R&D institutions within the country to develop S&T results in the least encumbered fashion, which tends to motivate and to increase the dedication of scientists and technicians to their work. Such autonomy also encourages ties to the users rather than to government agencies or departments and facilitates the development of contractual relationships with state or private enterprises. Autonomy does not mean that there is no government funding—this could be done on an endowment basis or through an agreed fixed budgetary support; it merely means that the R&D institutions do not have to be restrained by competition directly for budgeted funds with other governmental units; if they must compete, they are likely to lose out. Autonomy also means that funding should increasingly come from the users. Where there are serious gaps between the capabilities of potential users and local government support, intergovernmental entities or aid organizations should be drawn on, and units such as the World Bank and the Inter-American Development Bank should be encouraged to finance R&D institutions on an ad interim basis. However, it should be recognized that once these institutions succeed in producing relevant results, the government agencies related to these programs are likely to wish to take them over, and a means of preventing this needs to be devised.

One area in which R&D institutions appear not to have placed the highest priority is that on rural problems of agri-industry at the village level and what is known as "cottage industry." There are levels of technology between traditional production methods and more sophisticated levels in areas such as cement, sugar, bricks, and ceramics, and a separate R&D program should focus on how to pick up intermediate technologies from around the world and apply them in rural settings.

Still another need in the developing countries is the preparation of managerial orientations that will appreciate and search out technical solutions to production and distribution problems. If such a receptivity exists at the user end, the pervasive gap in most countries of an inadequate or nonexistent extension service (technical service corps) would be more readily filled. Korea has formed a technical service corps composed of university professors and other researchers who go out into rural areas and help teach what is needed on the spot; they have learned from this exercise the inappropriateness of many of their ideas and thus sharpened their application of technology to local needs.

The absence of an effective extension service points up once again the need for links among all actors in the development of S&T programs. University projects and results need to be brought before governmental R&D institutes, government ministries, state enterprises, and private companies. The work of R&D institutes needs to be tied into the ministerial programs and the objectives of state and private enterprises. The results of ministerial research groups need to be distributed to industry. The affiliates of transnational corporations should be tied into local university and R&D institutions. And efforts to diffuse technology backward to suppliers and forward to customers and laterally to partners or even competitors should be encouraged by such foreign affiliates, as well as by local institutions. Such an extension service is a prime requisite in the process of assimilation; poor assimilation capabilities have been a major factor in slowing the absorption and raising the costs of technology in many of the developing countries.

CONTRIBUTIONS OF TRANSNATIONAL CORPORATIONS

Left to their own decisionmaking in response to organizational, attitudinal, and market signals, transnational corporations will establish R&D activities abroad only when there is a sufficient market pull that requires a differentiated produce *or* where the local inputs are so different that materials-testing and technical-service activities require developmental efforts and when the affiliate can generate enough profit to support a local R&D activity. At least two of these requisites must be found in combination—the requisite funds plus one of the others. If these two exist, a foreign affiliate is a candidate for a development lab of an embryonic sort to undertake specific tasks. These tasks will likely be related to the domestic market or production processes, rather than being a part of a worldwide network, as shown in Chapter 1.

Where this pattern has not been followed in the initiation of overseas R&D activities, it is generally the case that the affiliate was purchased and already had an R&D unit in place. In other words, the TNC acquired a company with R&D laboratories, which it then melded into its worldwide network. When this has been the case, sometimes the additional labs have been given specialized programs to undertake or have been focused on unusual problems. In other words, R&D networks seldom permit duplication in effort among laboratories simply because the cost is too great and there is too much else that needs to be done.

Therefore the effort by host governments to encourage TNCs to establish highly specialized laboratories or set up a regional laboratory serving several countries faces considerable obstacles. These obstacles arise not only out of company practice and orientations but also from the difficulty of justifying either type in the absence of market pull. Even if there is a sufficient market to support it, a regional laboratory may not be justified because of difficulties of cooperating across national boundaries among many of the developing countries. National jealousies arise when such efforts are made at the R&D level so that considerable prenegotiation would be needed, probably among governments, to make this a feasible alternative.

However, TNCs can be encouraged by mere identification of needs to assist in the establishment and operations of private or governmental institutes, the improvement of university programs (especially to enhance their industrially related research activities), and in the formation of national technology centers that would assist in the diffusion of technology among various industry sectors.

Some of the international companies are already working with local governmental or private institutes—helping to train technicians, providing contract research opportunities, assisting in the development of standards, and so on. Some are also providing scholarships for graduate students in engineering and other disciplines, providing scientists as adjunct professors or occasional lecturers, assisting as advisers on program development, financing exchanges of professors so that they can see the way in which graduate programs operate at other universities abroad. Some are also providing research and bench experience in their own laboratories for graduate students or professors and are engaging local professors as consultants in their R&D projects.

CONCLUSIONS

The gestation of R&D activities locally as well as the successful transfer of technology require the development of an indigenous S&T infrastructure through the cooperation of at least three major institutions in any given country: governments, universities, and business. The way in which they will cooperate is determined in part by market forces (guided by either signals of supply and demand in the private sector of governmental demand) and the make-up of the science community that developed out of the interaction of these three major institutions. The linkages and the processes of development have been reviewed briefly in this chapter, showing that initiatives must be taken within each to compose a significant contribution, and cooperative efforts must also be instituted to link the programs of each with the other two.

But it must be stressed that the usefulness of all of these efforts depends on their contribution to the desired economic development goals in the country concerned. Therefore R&D and technology generation must be kept close to the users and to the final beneficiaries, rather than letting it be the hand-maiden of mere conceptualizers. Those who would establish complex networks without ties into production are likely to erect costly structures with very few developmental benefits. Equally, stress on adoption of technology to accelerate growth without the development of an appropriate local infrastructure will accelerate the use of foreign technology since it is readily available and proven out.

Therefore to get S&T policy oriented appropriately and moving in a useful direction requires simultaneous education of the industrial and consumer world of the specific applications of new technologies and their usefulness in meeting developmental goals and creation of infrastructure needed to supply the technologies desired. To build the infrastructure without the demand-pull is to raise costs; to create the demand-pull without the infrastructure is to raise expectations that cannot be fulfilled or that can be fulfilled only through increasing reliance on foreign technologies. To achieve the appropriate balance and pacing will require the application of the best scientific, technical, and managerial minds in both the public and private sectors.

NOTES

1. The study is based on personal visits over several days in each country, during which interviews were conducted with six to twelve S&T officials arranged by the U.S. Embassy in each country with ministries, laboratories, and various institutes. Each visit also turned up numerous materials that were useful later in writing appropriate descriptions. Each official was interviewed "without attribution," and none was given the opportunity to review the record.
2. See, for example, Sumitro Djojohadikusumo, *Science Resources and Development*, Jakarta, Indonesia; The Institute for Economic and Social Research, Education and Information (LP3ES), 1977; Michael J. Moravcsik, *Science Development: The Building of Science in Less Developed Countries*, Bloomington, Ind.: PASITAM, 1975; National Council for Science and Technology, *National Indicative Plan for Science and Technology*, Mexico, 1976; Arab Republic of Egypt, *Academy of Scientific Research and Technology*, Cairo, December 1976; Presidency of the Republic, *II PBDCT: II Basic Plan for Scientific and Technological Development*, Brasilia, 1976; and Council of Scientific and Industrial Research, *CSIR Handbook 1977*, New Delhi, India, 1977.
3. Moravcsik, *Science Development*, chs. 2–4.

Chapter 4

U.S. Responses to Needs of Science and Technology in Industrial Development

The U.N. Vienna Conference on Science and Technology in Development (1979) agreed on an action program to strengthen indigenous S&T capabilities of LDCs, to facilitate transfers of technology from advanced countries, to set up an information-retrieval network, and to establish a U.N. S&T capability. The research reported in the foregoing chapters was undertaken to help in the formation of U.S. responses to requests for assistance in strengthening local S&T in LDCs and in technology transfer. Several recommendations for follow-up of UNCSTD emerge from the research.

First, the U.S. approach should *not* be one of telling the developing countries how they ought to go about generating indigenous capabilities or expanding and accelerating the inflow of technologies from other countries. These countries are increasingly able and eager to establish their own policies, and they are gradually reassessing their ties to the advanced countries. The United States should adopt and express strongly its willingness to discuss the issues on a basis of partnership, mutual interest, and long-term commitment.

Second, it should emphasize bilateral negotiations and relations rather than multilateral, regional, or international (U.N.) arrangements. The problems faced by the developing countries are almost wholly internal in the sense that what they are seeking to do is to change domestic activities. There is little that needs to be done on an international basis, though what is done domestically will have international impact. Most of the countries are concerned primarily with their own development, and even though they are also concerned with regional development, it is more with the structure of industry and trade rather than the underlying science or technology—in the sense that the S&T used does not have to be, and is not likely to be, developed on a regional basis or out of regional or international institutions.

Third, the U.S. government needs to determine the size and nature of its commitment to assistance in the area of science and technology, establishing the total funds and manpower and private resources to be drawn on. It then needs to determine in bilateral discussions with countries the commitment that each of the recipient countries can make, shown by the dedication of top officials to the S&T objectives and the resources that they will commit. Once these commitments have been determined, specific programs should be decided from among the alternatives discussed previously with a schedule of performance laid out and provision for periodic evaluation of progress.

The objectives of such bilateral programs should be to assist in utilizing S&T more effectively in the pursuit of economic and social goals of the recipient country. This requires that these goals themselves be set out carefully and in some detail—at least in the form of indicative planning—else the S&T infrastructure will be built for one type of industrial growth despite the fact industrial development is moving in a different direction. Once these goals are determined, more specific programs of assistance can be detailed, helping to create the appropriate S&T capabilities and technology inflows.

Specifically, the programs undertaken should be directed to assisting in the creation of an indigenous S&T capability in the recipient country, toward creating appropriate demand-pull to get the technologies utilized in the industrial sectors needed to support the national objectives, toward establishing the necessary linkages for the transfer of technologies both internally and from abroad, toward establishing the necessary scientific infrastructure in educational and R&D institutions, and toward popularizing science and scientific inquiry to inculcate a scientific orientation in the early stages of education and to get potential users to look for means of applying S&T to their problems in both urban and rural areas.

In promoting the development of LDCs through the use of science and technology, the U.S. government must recognize that it is assisting in the transformation of these countries into more productive industrial units. It must therefore include in its own objectives toward S&T a greater ability to adjust to the industrial shifts among the LDCs. The LDCs will be picking up some of the industrial growth in specific sectors, exporting the product back to the advanced countries to be able to pay for the capital, management, and technology obtained. They will be attracting some industrial sectors into their countries, substituting for production existing in the advanced countries; they will be identifying segments of industrial sectors (products and components) that they can pick up more efficiently than is presently in the advanced countries; and they will be giving back industrial segments or sectors in which technology has moved on to more advanced states in the industrialized countries.

In U.S. government policy, a greater recognition needs to be given to the movements of and within industrial sectors that are caused by the transfers of technology among different echelons of countries. For example, some industries have moved into Japan and then from Japan to Taiwan and Singapore and will then be moving into Malaysia, Indonesia, Thailand, and other countries. Greater attention needs to be given among the OECD countries to the formation of industrial policies looking to smoother adjustments to industrial shifts around the world, and generating internal adjustment mechanisms, which should be developed with labor union support. Unless adjustment policies are adopted and are successful, increasing proposals will be pressed forward to slow the S&T growth in developing countries and consequent industrial changes. This can be only counterproductive to the reduction of tensions in the world. Consequently, a high priority needs to be given to this policy shift in the U.S. government—that is, from aggregate trade and financial movements to developments in specific industrial sectors. A policy position that indicated to the developing countries that the United States and the other OECD members recognized the necessity to come to grips with shifts in specific industrial sectors around the world would open up a much more cooperative dialogue since it would be seen that the critical problems were being grasped directly and firmly.

More specifically, there are a dozen ways in which the U.S. government can assist the developing countries in the formation and implementation of their own S&T policies. It needs to indicate that it would be willing to receive requests from them that would trigger assistance in development of indigenous R&D institutions (as done in the formation of the Institute for Scientific and Technical Cooperation), preparation of adequate engineering schools, development of business schools for better management, the formation of technology centers, the creation of extension services in S&T, encouragement of TNCs to cooperate in the diffusion of technology, provision of negotiation-simulation sessions, formation of dialogues on continuing problems, encouragement of international financial institutions to support R&D units in developing countries, and creation of data banks. In addition, in support of these activities, the U.S. government needs to reorder its organizational set-up for implementation, and it should promote the establishment of an International Industrialization Institute that would eventually help countries in the formation of their industrial goals and structures.

STRENGTHENING INDIGENOUS R&D INSTITUTIONS

To achieve self-reliance, the creation of indigenous R&D capabilities requires the establishment of autonomous institutions, separate from the ministries of governments to permit pursuit of programs and projects that are needed by the industrial sector. To achieve this autonomy, maintain adequate governmental funding, and gather the necessary resources requires a strong leader from the science field. He must also have the confidence of top officials in the government. Therefore one of the first moves in this activity is to assess the strength and leadership of the individual who will be responsible for the program in the recipient. The success of the U.S. response will depend on his abilities.

Once this person has been appointed, assistance can be given in the conceptualization of the structure of the R&D institutions— their place in overall governmental objectives, their orientation to government departments and to industry (both public and private), the scope of their activities, manpower needs, and so on.

Assistance will be needed also in research management—not only in providing some interim managers but also in helping to train institute managers, program managers, project supervisors, as needed in the growth of the institute. This training should be done largely in the States with independent institutes, company laboratories, and government labs on an exchange basis.

Assistance will also be required in the determination of appropriate programs—what industry sectors to emphasize within the overall industrial program of the government, the levels of technology to be worked on, the kinds of recipients to be served, and the inputs to be expected from university laboratories. More specifically, projects will have to be decided on and resources obtained, including attraction of national scientists who have remained abroad after training. Assistance will be needed in identifying these professionals and in recruiting them to return.

Assistance will also be needed in determining the precise ways in which the R&D institutes should be related to governmental ministries and to industry. For example, is the institute to undertake its own program and attempt to sell the results to potential users? Or will it establish advisory or governing councils composed of potential users so that the program is already designed to meet potential needs? Or will it be developed wholly out of response to specific requests from government and industry users? Although these approaches are not necessarily exclusive of each other, an assessment is required of what is feasible and desired.

Finally, assistance will be needed in the funding of these facilities since many governments simply do not have sufficient funds to initiate such autonomous institutions. They may become self-funding later, but initially a commitment of financial resources is required. To help support such an initiative on the part of developing country governments, the U.S. government should set aside 5 to 10 percent of its foreign-aid funds for assistance in the area of S&T, drawing on these for bilateral support of countries that take appropriate initiatives. A significant effort in a country starting from scratch could be mounted with around $15,000,000, encompassing a several-year start-up program.

The administrative structure for such assistance should be contracted out by AID (or ISTC) to a unit such as the National Research Council with its Academy of Science and Academy of Engineering, who could draw on their membership for appropriate exchange of personnel and for the types of assistance indicated previously. It would not be sensible to duplicate in AID capabilities which already exist outside, and AID (ISTC) could maintain the responsibility for determination of which countries are to receive bilateral assistance and the total amount of funding, then contracting out the actual implementation of that program. Provisions should be made for evaluation of the program so that both the NRC and AID/ISTC learn the lessons from experience and are able to pass them on to other countries through bilateral programs or a mere publication of the results of evaluations.

GENERATING INDUSTRIAL RESEARCH ORIENTATIONS IN ENGINEERING SCHOOLS

In developing manpower needed for the national R&D institutions—whether public or private—the largest resource must be developed locally, though some can be attracted from abroad. In developing this resource, local institutions will need assistance in developing appropriate programs and orientations. Technical training will be needed primarily within engineering schools but also within technical institutes. As in Korea, it may be necessary to set up entirely new institutes in order to avoid the delay of trying to change long-established curricula. Although education in the United States may be helpful in inseminating industrial-research orientations in a number of students and in training professors for engineering schools in the developing countries, this contribution is likely to be only marginal compared to the manpower that will be needed for all S&T activities in the developing country. In addition, the curricula and orientations of U.S. engineering schools are such that if a foreign student excells, his capabilities are oriented toward employment in the United States itself, and he is frequently attracted into that arena. Consequently, to serve the developing countries best, a primary effort needs to be mounted to generate national capabilities within engineering schools of the developing countries themselves.

These schools will need assistance in preparing their own professors, curriculum development, design of appropriate research activities, equipping laboratories, and placing graduates appropriately within the country itself.

The U.S. government should reinstitute and expand its effort under the so-called 211(d) grants extended in the early 1970s. These grants would assist U.S. engineering schools to create centers of expertise more relevant to the developing countries. Since the developing countries will insist on sending people into the United States, frequently with their own funds, they will need to be served better than they have been in the past. These programs should prepare them to return rather than induce them to remain in the United States. These funds should also stimulate U.S. engineering schools to undertake field research in developing nations to understand the needs and to develop their own curricula better. Specific government grants for faculty research abroad on technology needs would both assist in the development of national R&D institutions abroad and the reorientation of U.S. engineering curricula.

There are two administrative techniques for handling this type of program: One is to contract it to a specific engineering school that would work with others in arranging the exchanges and the seconding of faculty and personnel to the foreign schools. Obviously, if a completely new institute had to be set up in the LDC, the program would be longer and require more administrative resources. A second would be to contract the program to the National Academy of Engineering, which would then work through its membership and the established universities.

Unfortunately, the universities that are most likely to be useful to the developing countries do not have the highest international reputation, and some persuasion would probably be needed to bring the foreign professors and students to the most appropriate locations—which are frequently the land-grant technical universities—since they have closer ties to their users, and programs are at a level more appropriate to the developing countries than would be institutions such as M.I.T., Cal Tech, and Stanford. Yet the more prestigious schools are the ones to which developing countries wish to send their students or professors. A firmly established program that was able to demonstrate the appropriateness of other engineering programs would be more successful than simply handing out bilateral grants.

An ancillary part of this program would be the placement of engineering students being educated in the United States within companies that had affiliates in their home country to provide immediate employment opportunities back home. This placement could be accomplished during the summer or other semester breaks to provide an industrial orientation to the engineering curriculum.

One of the programs with which foreign engineering schools should become familiar is that undertaken by the Georgia Institute of Technology's Engineering Experiment Station (EES). This program is dedicated to assisting small rural industry within the state on an extension basis, and has been quite successful. The university has also extended this program into Ghana where it helped develop a Technology Consultancy Centre. This kind of activity should be expanded.

DEVELOPMENT OF INNOVATIVE
ATTITUDES IN MANAGEMENT

One of the problems in the developing countries is that there is insufficient demand-pull on technology, with much of the demand directed to external technologies, tried and proven. Managements in many developing countries are risk averse and are unwilling to accept locally generated technologies or even to try new technologies from any source whatever. It is likely that more formal management training would produce greater confidence on the part of managers and also open their eyes to the desirability of technological change and innovation in production and marketing. Assistance should be provided therefore, but only on request, for the development of curricula in business schools oriented toward inculcation of innovative attitudes on the part of young managers.

This type of assistance can be given only on the spot in the developing country. It is not particularly useful for professors to come to the United States to examine curricula over an extended time simply because the curricula in the United States are aimed at the needs here—not those in the developing world. They do not start with students with similar backgrounds nor in similar cultures, nor are they looking to similar end products. Many managers in developing countries should be oriented to small- and medium-sized industry, and the innovation there is of a different type from that in large-scale industry, which in any case is often state owned or affiliated with a transnational corporation.

Assistance from one business school in the United States to a similar school abroad has been useful in the past, but some have had disappointing results partly because there is no extended learning experience from one school to the next, and the learning curve is not accelerated. Each experience starts on its own and proceeds at a fairly slow pace. Therefore some central coordination would be desirable with an allocation of resources across the bilateral programs undertaken by AID/ISTC. This coordinating unit would mobilize the individuals to go abroad, provide them an orientation program based on experience in the past, provide basic program for discussion abroad, and assist them in the development of effective, least-cost administrative techniques, which are desperately needed in the business schools overseas.

The problem in the past in developing bilateral relations between U.S. and foreign business schools has been the inability or unwillingness of foreign institutions to alter their curriculum, which is partly a reflection of the inability of the professors to adapt to new directions or materials. Therefore a careful selection must be made of the readiness of specific countries to receive this kind of assistance. Unfortunately, those that are most ready to profit from such assistance are those that are most able to pay themselves, not needing foreign funding. Those that would need the greatest assistance are less likely to be able to assimilate it. Therefore very careful programming will be required.

This program also should be contracted out and preferably to a single university business school, which would then draw on faculties and help set up exchange programs among others. Alternatively, it could be contracted to the American Association of Collegiate Schools of Business, which would then have to build an administrative staff. It would be useful to help set up associations of collegiate schools of business in developing regions such as has been done in Latin America already. This helps to support the need for changes and to stimulate the specific changes required.

ESTABLISHMENT OF
TECHNOLOGY CENTERS

The purpose of setting up technology centers in developing countries is to support the users of technology, assist in technology assessment, and help develop the capabilities of assimilating technology, be it foreign or domestic. There are many technologies that are publicly available and could be applied locally if they were but known. A technology center should be a recipient of such information and be able to place the technology where it is likely to be used. This is a role that would not necessarily be played by R&D institutions, which are more concerned with developing or adapting technologies than simply being a channel for its transmission from one point to another, particularly if the technology is already publicly available.

A second objective of the center would be to train technicians in the repair and maintenance of equipment and in quality-control techniques. Many companies are not as productive as they should be because equipment has been allowed to deteriorate and cannot be appropriately repaired or maintained. This situation is due to a lack of technical expertise among the workers. In addition, even if some companies wish to establish higher quality standards, they would not be able to do so for a lack of technicians. Therefore the technology centers should be a means of disseminating appropriate standards through training technicians to understand their need and ways of achieving and maintaining standards.

The technology center should also provide repair facilities. And it should help inculcate quality-control techniques and programs. The center might even second technical personnel to companies to help set up quality-control procedures and a maintenance staff.

The centers could obtain foreign assistance through individuals such as those exchanged through the International Executive Service Corporation, thereby reducing the drain on local manpower. It should also be able to obtain training personnel from affiliates of transnational corporations in the country itself.

The center itself should be an adjunct to the extension services mentioned below and channels for influencing companies on the usefulness of following industrial standards, promulgated by the Institute of Standards and Measures. These departments or bureaus are being instituted in developing countries where they do not already exist, and increased attention is being paid to the use of standards especially in exports. Therefore they need to be applied to domestic production as well, and these technology centers should be a means of persuasion, coupled with an ability to repair and maintain equipment for testing and manufacturing so that the standards can be met.

The centers should be formed through joint sponsorship between the host country and the U.S. government, just as the agricultural "Servicios" were conducted after World War II in Latin America. The host government would provide the largest part of the funding since local facilities in the form of a building and manpower would be the predominant needs. The U.S. government could provide professional and technical manpower and encourage—through grants if necessary—the formation of a Technology Corps within the International Executive Service Corporation and a seconding of individuals from transnational corporations for a short term. The seconding could be stimulated by tax incentives if necessary.

The administration of the centers would be local, with the responsibility of the center extending over specific regions of the country—there may have to be several such centers. The program of the centers would be determined by consultation with potential users—local industry who would eventually be expected to make some contribution to the costs of the center. For example, the center might be used in the negotiations on a licensing agreement, receiving a fee for its consulting services. The center might also become a source of minor technical service under contract or fee from individual companies.

United States government funding of the direct costs of the center might be restricted to the provision of necessary equipment for repair facilities and instruction on maintenance of various types of machinery. In order not to have the centers duplicating a lot of machinery, different centers should be specialized in particular industrial sectors.

CREATION OF EXTENSION SERVICES

As mentioned with reference to the technology centers and also with reference to the engineering school programs, extension services are needed to reach industrial users so that they do in fact adopt and learn how to assimilate technology in the manufacturing and commercialization processes. Although extension services in agriculture are fairly readily understood, the need for extension services in industry is not quite so clear, given the fact that they have not existed in a formal sense in the advanced countries. However, there are numerous routes through which technology can flow to companies seeking it. Among the advanced countries, there has been a flow through published materials, competitive action, conferences of many sorts, personal contacts in an industry sector, and through licensing and other intercompany relationships. In developing countries where there are few competitors in a given sector, little indigenous technology, and few published sources, primary reliance has been on foreign sources that provided their own extension services through technical assistance agreements.

Therefore these types of services will have to be generated de novo in developing countries. The effort by Georgia Tech mentioned before can be a prototype, matched by the experience in Ghana. The extension services should be the link between the sources of technology and the ultimate users and are of course best formed by those who also develop, assess, or apply the technology itself. In other words, such services should be located in or tied in with agricultural and engineering schools or public R&D institutes or companies. Responsibility for making sure that the extension services are developed and operate appropriately should be housed in one agency, which could be within the technology centers since they will also have lines into the industrial users. However, the expertise needed for technical services related to the adoption, assimilation, and implementation of technology is different from the program proposed previously for the technology centers. A different training is required for this objective compared to repair and maintenance, for example. The technique of seconding technicians to the companies for a short term is the same, but the work that would be done would be substantially different.

The U.S. program of assistance to extension services should consist of the provision of training certain professionals, the development of materials to assist in teaching companies how to select, assimilate, and implement technologies brought in; the provision of short-term personnel to assist in setting up the services; and training in administration of the programs. Once again, this should not be a costly program but one requiring substantial administrative capabilities and experience with these kinds of activities. For this reason, the program should be contracted out and probably to an entity such as Georgia Tech, which has had some direct experience. AID/ISTC would negotiate an overall agreement with the developing country that would include the establishment of extension services, with this part of the program being contracted out for implementation.

STRENGTHENING NEGOTIATION CAPABILITIES

Despite the existence of some private and international programs aimed at helping strengthen the negotiating capabilities of developing countries, there is still a substantial need for this kind of assistance. The U.N. Centre on Transnational Corporations provides two- to four-week courses for governmental officials of a single country covering a range of subjects related to negotiation and the formation of policies on foreign investment and technology transfers. However, it can reach only six or eight countries a year, and the need is much greater. Most of the experience in negotiation is on the side of the companies since they have been working with a number of different governments and with a variety of companies in direct negotiations. Consequently, governments are continuing to seek ways of improving their bargaining positions and their bargaining techniques. One of the advantages of developing seminars on the subject is that it forces an assessment of the bargaining position of the recipient of the country, which permits an analysis of the pros and cons or the attractiveness of the country as a market that would pull in investment or as a system that could assimilate technologies. Also by the mere process of going through particular exercises in negotiation, additional information will be gathered on sources of technologies and problems in the process of technology transfer— which of course would be the subject of negotiation. Still another benefit from this process is improving the communication among government departments within the developing country, as each sends individuals to the negotiating session and they begin to understand only the common problems they face but the different views that they hold within their own ministries. Finally, a process of simulation of negotiations helps to clarify the issues that are likely to be faced.

In setting up negotiation-simulation seminars, it is important to focus the dialogue on specific kinds of situations. Therefore it is best to hold these sessions with a single government and have a single industrial sector as the subject matter. For example, negotiations on investment and technology transfers in the mineral sector would lead to considerations different from those relating to either automobiles or pharmaceuticals. To get into the specific issues, it is best not to mix the various sectors. Consequently, several sessions would be needed to help a government understand the range of problems that it faces in negotiation and the techniques and procedures available to it.

In setting up such seminars, one needs to have as much experience from outside the country as possible as well as the largest number of individuals from the local government who are likely to be involved in negotiations on a particular set of issues. Therefore participants would include several from, say, the Ministry of Energy, plus some from the Ministry of Finance and the Ministry of Labor, faced by a representative of the industrial community—both home and foreign; plus international institutions such as the World Bank and UNIDO, plus some legal counsel. These individuals would be guided by private consultants who are trained in the process and negotiations simulation.

The dialogue within the seminars would cover a range of problems such as the strength and weaknesses of the bargaining position of the host country, constraints on the entry of direct investment, differences in licensing agreements, origins of technology apart from investment, concession provisions, management contracts, problems of renegotiation, problems of closing and termination of agreements; and specific contract provisions relating to matters such as control, ownership, remittance, capital repatriation, property rights, duration of the agreement, governmental incentives, dispute settlement, termination, and any adjustment cost to renegotiation.

The host government should be willing to bear a substantial cost of such seminars, mostly in the provision of a local facility plus the expenses of participants. External costs such as travel and fees for consultants running the seminars could be met from a variety of sponsoring sources such as private foundations, transnational corporations, the World Bank, and USAID/State.

Groups that might cooperate in the sponsorship of such seminars would include: UNIDO, UNDP, the UN Centre on Transnational Corporations, OECD, ADELA, PICA, National Chambers of Commerce, Industry Associations, the Asian Development Bank, Inter-American Development Bank, and local business schools.

Although it is not difficult to find *some* place to hold a meeting of this sort, appropriate facilities are not readily available in every country. Facilities do exist in the Philippines and Singapore, and Seoul, and a few places in India. Kenya and Nigeria have facilities, as do many countries in Latin America.

However, the largest bottleneck is in finding appropriate personnel to conduct such seminars and to prepare appropriate case materials for industry sectors that will be instructive to the various countries needing this type of training. The personnel would come from a small list of academics who have been working in this field internationally (unfortunately, only eight or ten such professors exist worldwide) and an additional small list of consultants and a few from companies who have been directly involved in negotiations and would be permitted by their companies to join such sessions.

Again not a large amount of funding is required, but a substantial amount of program effort is needed for such a continuing activity. To put on even thirty such sessions a year, which would still not meet the need, would require a substantial administrative effort. Such a volume of activity would probably run into the bottleneck of appropriate personnel to conduct the seminars.

The *external* funding per seminar for a two-week period, covering travel per diem, and fees would probably amount to no more than $10,000, plus administrative costs averaging $5,000 per seminar. A total program of thirty such sessions would therefore require a contribution of $450,000 per year over, say, a five-year period—after which others could be expected to pick up the continuing flow of assistance. Costs of the program would be contained if it were held on a bilateral basis; regional programs require extensive travel on the part of the participants themselves.

A group such as the Fund for Multinational Management Education would have the necessary entrées into academic resources, company officials, and consultants, plus contacts with government officials that would permit the conduct of such programs effectively. It has already demonstrated the feasibility of such programs in several countries, but the bottleneck for expanding the activity is the lack of external funds. The Georgetown University Centre has also conducted a few such seminars; and given the need, its resources could be expanded.

ESTABLISHMENT OF
CONTINUING DIALOGUES

Since the Vienna Conference did not encompass setting country priorities but mainly focused attention on the significance of S&T policies and their contribution to economic development, continuing dialogues are needed. Fora should be sought for communication of cooperative objectives and mechanisms. These fora should be located in the developing countries to obtain the largest effective participation from officials there and to reduce their costs, thereby increasing the likelihood of requests to establish such dialogues.

These dialogues should not be similar to the negotiation-simulation seminars but rather involve officials (just below ministers) who are concerned with a broad range of policies concerning the use of S&T in economic development, beginning with the formation and specification of development goals themselves. The participants should basically be technically oriented (career) personnel, involved in the process of applying technology to developmental objectives and developing the S&T infrastructure. They should not be the political authorities, although these will obviously need to be involved at some juncture and approve any recommendations.

A fairly wide range of subjects would need to be covered to respond to the program agreed on in Vienna. In the area of industrial goals, the following seem significant enough to warrant an effort to establish a continuing fora on a bilateral basis:

1. The criteria of assessment of appropriate technology that should or would be used by the host country.
2. Concerns over provisions of technology agreements.
3. Maintenance of confidentiality of the data transferred under agreements.
4. Protection of proprietary industrial property.
5. Regulations concerning duration of visits of technicians from abroad.
6. The needs for continuing technical assistance.
7. Requirements that recipients of technology export products.
8. Guarantee requirements concerning technology transfers.
9. Problems of renegotiation of contracts.
10. Arbitration or settlement of disputes.
11. Taxation issues, both host and United States.

12. Possibility for contracting of training for better assimilation of technologies.
13. Coordination of ministerial policies within the host country.
14. Processes for establishing linkages among foreign R&D activities of affiliates of international companies with R&D capabilities in the host country.
15. Potential contributions of newly industrialized countries to lesser developed countries, through both technology generation and technology transfers.
16. Improvement of educational curricula.
17. Stimulation of entrepreneurial attitudes.
18. Popularization of science and technology.

The desirability of such dialogues has been established, as has the fact that they can be successful, drawing the appropriate people together for one or two days. These dialogues would not only help focus the issues but would lead to the establishment of appropriate cooperative mechanisms on a case-by-case basis. Out of such dialogues are likely to come a series of significant but not so costly moves to facilitate the development and utilization of S&T in the developing countries as well as direct contributions out of the pool of S&T in the advanced countries.

Since the participants in these dialogues should include industry officials, professionals from R&D institutions, university officials, as well as representatives of the two governments, this program also can best be carried out by contracting to an independent entity that is tied in with all of these groups. Multiple sponsorship should be sought so that it is not the initiative of any one group and so that it does not appear to be official, with the implication that the results are to be directly implemented by either government. It would be desirable for such dialogues to be institutionalized by a few countries to begin with, with sessions on the order of once every three or four months. Experience would soon indicate whether these sessions needed to be that frequent and whether eventually the sponsorship could be left to the other country, permitting it to take the initiative even in the establishment of the schedule, program, and participants, leaving the U.S. government to fund only its own participation.

Initially, program funds should be made available on the order of $500,000, which should be sufficient to fund bilateral fora with ten or twelve countries, including the preparation of appropriate background or discussion papers.

CREATION OF TECHNOLOGY
DATA BANKS

One of the significant gaps in the capabilities in developing countries to formulate technology policies is the lack of information as to available technologies already existing in the world. Two issues surround availability—the existence of the technology and access to it by a developing country. The sources of technology that should be available to developing countries are several, both from within the country and outside: publicly available information, private information that is nevertheless nonproprietary, proprietary information, and government-created technologies. Not all the information or technologies created by these groups will in fact be accessible or even in use. When it is not in use, little information will be disseminated. It is therefore necessary to find ways of gathering information on technologies that exist, whether or not they have been tried.

The U.S. government should assist countries on a bilateral basis on setting up data banks and access to data banks in other countries—including that in the United States. In establishing a national data bank, of most critical importance is what to put in it. This requires determination as to the areas of interest of the country and then a means of finding the information to process. The second question is how to process it so that it is retrievable against the inquiries that are likely to be made. A third is to whom the information in the bank is to be available—should there be a clearing-house procedure for screening requests, refining them and making them more precise in obtaining the information from the bank. For example, many potential users of technology do not even know what it is they might be looking for and do not know how to frame appropriate questions. This may require an interface such as a technology center where a potential user might be stimulated to ask relevant questions. The retrieval process is therefore quite important; otherwise information in the bank will not be used to respond to questions that would be asked.

There are many proposals for international data banks and even for regional data banks. However, it is likely that an international data bank will be so extensive and so difficult to program the information in and to retrieve it that it will not serve the purposes of the developing countries very well. Information to be used effectively needs to be as close to the user as possible; whereas an international data bank is as far away as possible. Only if the retrieval process is close to the user can the significance of the information and the best use of it be determined. The process of dissemination of the information throughout the system is critical, which requires an understanding of the environment in which it is to be used. Several officials interviewed questioned whether even a national data bank—unless carefully segmented to reflect the needs of different sectors—would be close enough to the users to be effective.

While it will be useful for the U.S. government to gather information on technologies available in the United States, the next step would seem to be to disseminate that information into the national and local data banks where it can be close to potential users. The establishment of national and local data banks would develop an expertise in the developing countries on identification of technologies as well as identification of potential users.

The U.S. government should allocate funds to pay the external costs of collection and transfer of information, which will be needed even to gather information that is government owned, publicly available, or privately owned but nonproprietary. Proprietary information is by far the smallest portion of all technology information useful in developing countries, probably amounting to no more than 20 percent of the total, and companies could still use the data banks to put sufficient information in for the stimulation of preliminary talks with potential users.

The administration of the data bank should be in the hands of the developing country, and U.S. contributions to these costs should be only for the external expenses, under a bilateral program agreement. This part of the program could be administered by the Department of Commerce through its National Technical Information Service.

FINANCING BY
INTERNATIONAL INSTITUTIONS

To expand the funds available for the preceding programs, the U.S. government should be prepared to cooperate with initiatives from developing countries to encourage the international financial institutions (World Bank, International Finance Corporation, Inter-American Bank, Asia Development Bank, etc.) to allocate funds for the support of S&T infrastructure. These institutions might well provide funding supplementary to that of the U.S. government and other countries for the gestation of R&D institutions in particular countries, possibly taking responsibility for certain sectors such as agriculture or minerals.

These institutions could also assume responsibility for helping develop technical expertise in the identification and storage of technical information so that data banks were more effective in the search and retrieval process.

Implementation of this proposal could be left to the U.S. Treasury, which is the agency responsible for participation in the international financial institutions.

ESTABLISHMENT OF AN INTERNATIONAL INDUSTRIALIZATION INSTITUTE

As emphasized in Chapter 3, the first requisite for an S&T policy is determination of industrial priorities. Few developing countries are set up to do this—conceptually or administratively. Even if they were, the information needed to avoid mistakes (excess capacity, high-cost facilities, etc.) is not readily available. And advice from advanced countries is not necessarily useful and may not fit with LDC goals. To help fill these gaps, formation of an International Industrialization Institute has been proposed.[1] It would be of considerable value in providing the information and analysis necessary to set goals, which S&T policy would help pursue and to define alternative technological routes to development. Creation of this institute could be achieved by one of three routes:

1. An international foundation could be established for the support of research on industrialization, which would extend grants under contract to existing research institutions or researchers. Contributions to the foundation could come from a wide range of private and governmental sources, with the results coming back to the foundation for worldwide dissemination.
2. A new international center for research on industrialization could be created as a private, autonomous but multinational organization, governed by a board of directors with international credentials. It would be separated on the political pressures with funding on a continuing or endowment basis. It would both conduct and sponsor research, again making the results available worldwide.
3. A research center could be established within an existing international institution such as the World Bank but wholly dedicated to the investigation of the process of industrializa-

tion. The only drawback to this approach is the subservience of the research to possible governmental interference or priorities that might shift the focus of research and interrupt it in ways that would make it less productive in the long run.

If either of the first two approaches is adopted, the government should turn to the National Research Council to implement the gestation of the institute, providing preliminary funding for this process. If the third approach is chosen, it should turn the proposal over to the U.S. Treasury for its insemination through the established processes of the World Bank.

*   *   *

There are two other activities in which the United States needs to engage in order to undertake the preceding programs. These require a U.S. initiative, but unilaterally and without discussion with the developing countries. Both are necessary for success of the programs: One is a procedure for encouraging transnational corporations to cooperate in the formation and implementation of S&T policies in developing countries, and the other is the reorganization of the U.S. government's responsibilities in S&T for development.

TNCS' ROLES

TNCs, with affiliates overseas, will follow a logical process of expanding their R&D activities, which is related strictly to the capabilities of the affiliate and market stimuli. Governments may attempt to accelerate this process, but TNCs may not respond appropriately. "Appropriately" does not necessarily mean doing what the host country requests but helping to form the relationships so that the requests are in fact effective and supportable by the companies themselves. These are actions that are mutually desirable, and the U.S. government could stimulate them merely by forming an effective dialogue with representatives of the companies themselves. Some of these actions include the following:

1. Establishment of quality-control and technical-service facilities in the developing countries.
2. Dissemination of R&D *attitudes* to suppliers and customers, helping them to develop a receptivity to new technologies, and encouraging diffusion among their competitors.
3. Establishment of "quality-control circles" within the indus-

trial sector to stimulate the adoption of governmental industrial standards or even higher ones if necessary for export.
4. Promotion of exchanges of personnel to raise productivity among suppliers and customers with complementary companies in the United States.

To provide some incentive for exchange of appropriate personnel to assist in some of these programs, the U.S. government could provide tax incentives when individuals are seconded to a host government or to independent companies in an industrial sector to improve technology assimilation. Similarly, seconding of company officials to work with engineering schools in developing appropriate curricula or the bringing of professors into industrial labs to inculcate attitudes of industrial research could be appropriate programs for subsidy.

The initiative of the U.S. government should be in the form of a convocation of R&D leaders in international companies to discuss the ways and means of adopting a more cooperative stance in the developing countries—on company initiatives. These leaders are largely members of the Industrial Research Institute, and the first step might well be a presentation of issues before an annual convention of the IRI, with subsequent discussions in small groups.

The government should also be prepared to finance the exchange of productivity teams such as was done with the European Productivity Teams under the Marshall Plan. This exchange could be contracted out for administrative purposes, as some of the programs suggested previously.

ORGANIZATION OF U.S.
GOVERNMENT EFFORTS

There are a number of entities involved in the problems of S&T in international development, the most recent of which are the proposed Institute for Scientific and Technical Cooperation and the Appropriate Technology Institute. Both were projected with specific reference to the developing countries. In addition, both the Department of State and AID have their bureaus and offices related to science and technology, as does the Department of Commerce. The White House Science Adviser has particular responsibility. And the National Science Foundation is sponsoring research on the subject.

As can be seen from the prior recommendations as to the U.S. government program, much of the operation of the bilateral programs can be contracted out. And the central responsibility for contracting can certainly be located within AID/ISTC. However, the establishment of a bilateral agreement with individual countries on S&T should encompass the activities of the other departments or agencies involved. Therefore a coordinating responsibility is required within the government—not only to coordinate U.S. government operations but also to coordinate with those contracted for implementation by agencies outside of the government.

To achieve this coordination, to stimulate cooperation from the companies themselves, and to involve the science community and labor interests, an advisory committee should be established that draws on representatives from each. Out of this group would come a number of participants for the various dialogues that were mentioned before.

Unless there is a centralized responsibility for following up on the Vienna conference, the various initiatives or programs are likely to be dispersed among existing units within the government, leaving little impact and little opportunity to evaluate what has been done even with a particular country. This central responsibility must be assigned in such a way that it is charged with implementing a coherent set of programs. Finally, there should be a responsibility for this unit's reporting periodically to the Science Adviser and to the Congress on progress under the program to determine the usefulness of separate parts and assess progress. Efforts should be made in this evaluation to indicate parts of the program that can be transferred into the private sector or simply permitted to die once they have succeeded or the emphasis shifted from one industrial sector to another or one geographic region to another. Not everything can be done simultaneously, so priorities will have to be set, which should be the responsibility of this central unit under the advice of the committee mentioned before.

With its own administrative structure in place, the U.S. government can turn to mobilizing its own S&T resources, and those of the private sector, to combine with those in developing countries to give sound support to industrial developments in LDCs.

NOTE

1. National Academy of Sciences and National Academy of Engineering, *Meeting the Challenge of Industrialization*, Washington, D.C., 1973.

Index

About the Authors

Jack N. Behrman is Luther Hodges Distinguished Professor at the University of North Carolina Graduate School of Business Administration. He has held faculty appointments at Davidson College, Washington and Lee University, and the University of Delaware, and visiting professorships at George Washington University and the Harvard Business School. In addition, Dr. Behrman is a frequent member of research panels for the National Academy of Science and the National Academy of Engineering; an advisor to the U.S. Department of State and the U.N. Centre on Transnational Corporations; and Senior Research Advisor to the Fund for Multinational Management Education in New York. From 1961 to 1964, he was Assistant Secretary of Commerce for Domestic and International Business. Dr. Behrman is the author of numerous articles, books, and monographs, including *Some Patterns in the Rise of the Multinational Enterprise* (1969), *National Interests and the Multinational Enterprise* (1970), *U.S. International Business and Governments* (1971), and *The Role of International Companies in Latin American Integration* (1972); he is coauthor of *International Business–Government Communications* (1975) and *Transfers of Manufacturing Technology Within Multinational Enterprises* (1976).

William A. Fischer received a B.S. (in civil engineering) and an M.S. (in industrial management) from Clarkson College and a D.B.A. from George Washington University. His work experience includes both industrial and government positions in the management of high-technology projects. At present, he is an assistant professor at the School of Business Administration at the University of North Carolina at Chapel Hill. Among Dr. Fischer's recent publications are articles on technology transfer, scientific and technical information and the performance of R&D groups, and postwar Japanese technological growth and innovation. He is a member of the editorial advisory board of the *Journal of Technological Transfer*. His research interests include the management of technological change, technology transfer, and corporate technological strategies.